Advanced Construction Project Management

Advanced Construction Project Management

The Complexity of Megaprojects

Christian Brockmann
Hochschule Bremen
Bremen, Germany

Registered Offices
John Wiley & Sons, Inc., 111 River Street, Hoboken, NJ 07030, USA
John Wiley & Sons Ltd, The Atrium, Southern Gate, Chichester, West Sussex, PO19 8SQ, UK

Editorial Office
9600 Garsington Road, Oxford, OX4 2DQ, UK

For details of our global editorial offices, customer services, and more information about Wiley products, visit us at www.wiley.com.

Wiley also publishes its books in a variety of electronic formats and by print-on-demand. Some content that appears in standard print versions of this book may not be available in other formats.

Library of Congress Cataloging-in-Publication Data

Names: Brockmann, Christian, 1954- author.
Title: Advanced construction project management : the complexity of megaprojects / Christian Brockmann, Hochschule Bremen, Bremen, Germany.
Description: Hoboken, NJ, USA : Wiley-Blackwell, 2021. | Includes bibliographical references and index.
Identifiers: LCCN 2020020092 (print) | LCCN 2020020093 (ebook) | ISBN 9781119554769 (hardback) | ISBN 9781119554745 (adobe pdf) | ISBN 9781119554752 (epub)
Subjects: LCSH: Engineering–Management. | Building–Superintendence.
Classification: LCC TA190 .B75 2020 (print) | LCC TA190 (ebook) | DDC 624.068–dc23
LC record available at https://lccn.loc.gov/2020020092
LC ebook record available at https://lccn.loc.gov/2020020093

Cover image: © Kyle Garrity/EyeEm/Getty Images
Cover design by Wiley

Set in 9.5/12.5pt STIXTwoText by SPi Global, Chennai, India
Printed and bound by CPI Group (UK) Ltd, Croydon, CR0 4YY

10 9 8 7 6 5 4 3 2 1

To those who strive to close the gap from man to man
By viewing culture as a gift and not a hassle;
To those who love to build a bridge from place to place
For mankind, culture, place and space go hand in hand.

Contents

1

Introduction

Construction megaprojects capture our imagination. They are the most visible and most lasting products of human ingenuity. Humankind has erected iconic buildings and statues across time in all parts of the world. Most of them were megaprojects at the time of construction. Although the examples shown in Figure 1.1 are in black and white, most readers will be able to recognize them. The chosen examples include monuments from the East to the West:

- Sydney Opera House (Sydney, Australia, 1973) is the best-known building in Australia
- Akashi Kaikyo Bridge (Kobe, Japan, 1998) is the longest spanning bridge in the world
- Oriental Pearl Tower (Shanghai, China, 1994) has a unique shape and is 468 m high
- Taj Mahal (Agra, India, 1653) is a masterpiece of Muslim architecture
- Burj Khalifa (Dubai, UAE, 2008) is the tallest building in the world
- Kremlin (Moscow, Russia, 1561); the St. Basil's Cathedral is a fine example of Russian Orthodox architecture
- Pyramids of Giza (Cairo, Egypt, 2560 BCE) are some of the oldest and most massive structures in the world
- Hagia Sophia (Istanbul, Turkey, 537) was built as a Byzantine church, and remodeled as a mosque after 1453
- Eiffel Tower (Paris, France, 1889) is a representative structure built with wrought iron
- The Statue of Liberty (New York, USA, 1786) is a symbol of and dedication to liberty
- Machu Picchu (Peru, fifteenth century) is a city built by the Incas
- The Moai statues (Easter Island, Chile, after 1200) are spiritual statues

The list is anything but exhaustive since I have omitted civil structures such as tunnels (Gotthard Base Tunnel, Switzerland), dams (Hoover Dam, USA; Three Gorges Dam, China), or roads (New Silk Road, Asia, Africa, and Europe) because it is not easy to depict such structures. No products from Airbus, Apple, Gucci, JP Morgan Chase, or Porsche can boast of the same imaginative power that these structures emanate.

Awe and attraction are not the only feelings we develop toward megaprojects. Many of them have met with outright rejection during planning and execution. Not everyone appreciates spending billions for a single project, and not everyone makes use of the same structures. Some people might enjoy a football match at the Wembley Stadium (completed 2007), while others prefer a visit to the Oslo Opera House (completed 2008). Nobody likes a nuclear power plant in the backyard or a highway in front of the house. All megaprojects

Advanced Construction Project Management: The Complexity of Megaprojects,
First Edition. Christian Brockmann.
© 2021 John Wiley & Sons Ltd. Published 2021 by John Wiley & Sons Ltd.

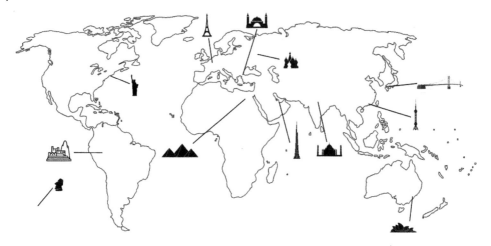

Figure 1.1 The world in iconic buildings and statues.

have positive and negative aspects; individuals can choose their own perspectives. It is not the task of civil engineers to make decisions for the implementation of megaprojects; this, in many countries, is a democratic procedure. However, each civil engineer must decide for himself whether to engage in a megaproject or not; this is his civil duty.

A case in point is the Elbphilharmonie in Hamburg designed by Swiss architects Herzog and de Meuron (Figure 1.2). Its scope changed considerably over time, construction costs increased more than tenfold, and construction time exceeded the initial plan by years. Reports in German and international publications were scathing during construction. Immediately after opening, journalists heaped praise over praise on the building. Founded on an historic storehouse in the port of Hamburg, the peaks of its roof soar into the sky. Hamburg became a wealthy city through trade, and its music transcends the borders of our world (Brahms was born in Hamburg). We can read this and much more into iconic buildings. The number of visitors at the Elbphilharmonie in 2017 equaled that of the Sistine Chapel in the Vatican in the same year. The experience is two-sided: not all criticisms of megaprojects turn to praise. However, many megaprojects are not only the dreams of civil engineers but of humankind.

Literature on megaprojects often concentrates on the front-end, stressing how important governance structures are. Construction itself seems a mere nuisance. This turns the reality upside down and I will call it the fallacy of retrospectivity, because authors often interpret earlier events based on information that becomes available much later. Construction without a plan for the front-end is not advisable. Planning without construction might be fun but not worth the expenditure. We should try to avoid decisive mistakes during planning and construction.

Planning and construction are like the two blades of scissors: they work together. One without the other is useless, and both are equally important. Overestimating the front-end is a serious mistake; underestimating it will also prove catastrophic in megaprojects. Given the abundant front-end literature, this book is for those who can hold a shovel and grasp complex ideas at the same time – people with a strong foundation in construction and ideas soaring into the skies just like the roof of the Elbphilharmonie.

Figure 1.2 The Elbphilharmonie in Hamburg. Source: sirius3001/123RF.

1.1 Let me Start with a Story

Imagine you meet Choi from South Korea at an international contractors' conference. He was the project director of the Burj Khalifa, the tallest building in the world with a height of 828 m. You see him discussing something with an American, Alex, who was the former project manager for a petrochemical plant in Taiwan. Also present is Khaled, responsible for the construction of Masdar City, projected to achieve zero carbon emissions while housing a population of 50 000. Khaled is from Syria and has worked in the United Arab Emirates for the past 15 years. Since you are standing next to them, you cannot help following their discussion. As it continues, you become more and more intrigued.

ALEX: *…so awfully proud when we opened the plant on time; it was pure exultation.*

CHOI (LAUGHING): *Yeah, yeah, I remember our celebration when we reached 828 m, it filled us with pride and we were cheering. After all the headaches we faced over the years, we finally finished our task. I guess it doesn't matter whether you break a record in height or in length or no record at all. The pride comes more from all the difficulties we had to overcome.*

ALEX: *Of course, I also remember all those sleepless nights, especially at the beginning when everything was chaos. It was a hard fight, finding solutions for all the technical and the management challenges. All the different stakeholders with all their different goals drove me mad at times. My boss was only concerned about the profit. There was no visit without him asking about profit, from the beginning to the end of the project. I can tell you we had a very tight budget. It was definitely a most demanding time. But man, I tell you, we did it! Yes, we did it!*

CHOI: *I agree, it was demanding in my case as well. However, sometimes I had the feeling I went bungee jumping every day and my company even paid me well for it; this was the adrenaline pumping from morning to evening. I simply cannot understand people who are bored by their work.*

KHALED: *I wish I could join your jubilation but I'm stuck in the middle of my project, and some days, I just try to keep my head above the water. I'm fighting all day long to get some structure and routine installed. There are thousands of loose ends, and I have no clear idea where to start and how to proceed. Of course, when my boss comes, I can't show my doubts; in my culture, you can never show weakness. Certainly, I'm never bored; I've reached my limits. To me, this is the bloody mother of all projects.*

CHOI: *...the adrenaline?*

KHALED: *...surely keeps pumping. If I would just know that I would somehow reach our goals. This insecurity when making decisions and this ambiguity in so many problems... I sometimes feel like I'm lost in the desert. There is not enough information, no clarity. Everything affects everything else, nothing is straightforward. There are so many iterative steps...*

ALEX: *I can understand and empathize with you. It does take a lot of patience and perseverance. Choi, do you also remember those endless days when nothing seemed to move?*

CHOI: *...and some days when I didn't want to leave the bed in the morning, just keep hiding under the blanket.*

KHALED: *Masdar City is not my first megaproject, of course. I've been involved in so many of them, I can't keep track. In the Middle East, there are so many of them. Yet, every megaproject is so very different. We are always facing new challenges, and there is no one recipe for success.*

CHOI: *I believe the most important characteristic for the top management is the ability to learn, to learn very quickly.*

KHALED: *I did not start my project; there was another project director in the beginning. He had a big ego and told everybody about his experience. He impressed our bosses by his self-assurance and he was quite a bigmouth. I believe he actually expected a sure path to success. He tried to do everything the same way as in his last project, which had been very successful. Only in our case the situation, the tasks, and the problems were quite different. He never understood that he had to find new ways; that he had to learn. It ended in disaster.*

ALEX: *I have seen that on many projects. Project leaders who did not understand the importance of learning. I mean, how can you think to use the same old tools for a completely new challenge? That's so stupid.*

CHOI: *Of course, it is easier to apply a tested and proven approach...*

ALEX: *...did you try it in your project?*

CHOI: *Yes, in the very beginning, but we soon found out that it does not work. We hit roadblocks every single step. It took time to find a way bypassing them. This is what I call learning, finding a way.*

KHALED: *In the Arab world, we like hierarchies. Therefore, when my bosses promoted me to the position of project director, they expected me to control everything. I went crazy working 16 hours every day, and still the unsolved problems kept piling up. I found out that I had to share the burden with colleagues. An old company friend was the construction manager and we have been working together for many years. We also come from the same area in Syria. I shifted some of the burden and decision-making to him. That was easy, as I trusted him.*

ALEX: *In megaprojects, hierarchies simply don't work. You have to delegate work, to share the decision-making. We have been working with lean management in the USA for quite some time, something we invented, I guess. There is too much dynamism in megaprojects for a strict hierarchy. Agility is what we need.*

CHOI: *...If you are not able to trust, I do not know how you can manage a megaproject...*

ALEX: *...yeah, you even have to trust complete strangers, since there are always new colleagues in every project. It's easy to trust someone you've known for many years, but I have problems with newcomers. However, there is no way out of it; the workload is too high.*

CHOI: *...Sharing the burden, delegating the work, and trusting others to perform well – they are indispensable. It is also tough because, in the end, I got the blame when something did not work out. My bosses kept to the traditional hierarchical thinking.*

ALEX: *...We were a team of six who made all the big decisions; everyone took smaller decisions by themselves. If you would ask me to describe how we worked, I would answer that all six of us were exploring different paths and venues, many of them leading to dead ends, maybe even most of them. Then, suddenly something works and you keep pushing ahead until there is another dead end and you have to take a turn again. Most of the time, our paths were divergent, and sometimes, they became convergent for a bit, only to drift apart again. Yet, after a while, a solution emerged that seemed promising. The disappointments became fewer and the successes more frequent. Indeed, a structure appeared, and we kept pushing it. I wish we could have known this structure in the beginning!*

CHOI: *As you said before, there is only chaos in the beginning. In the end, this is the fun part, to keep working on the chaos until it becomes structured and manageable. That is so very satisfying.*

KHALED: *I know this from other projects, but in my case, I can't see the end, I don't see the structure. We still do not have enough information, and when I have to make a decision, I feel quite inadequately prepared.*

CHOI: *You make a decision and, half a year later, someone will criticize you based on the information accessible at that time. They don't take into account the difference in information. It is so easy to predict the result of a football game after its end! Yes, we face all the problems we discussed, but, in the end, megaprojects can give you more joy and satisfaction than any other work that I can think of. They fill you with the pride of an astonishing accomplishment. In the end, I am a world champion. I have built the tallest structure in the world!*

ALEX: *Yeah, that's true!*

KHALED: *I am not so sure at the moment…*

The Korean name "Choi" means pinnacle. The name "Alex" is short for Alexander, and we know Alexander the Great for his conquests. Finally, "Khaled" means immortal. Choi, Alex, and Khaled are purely fictional characters, and they were in no way involved in the construction of the Burj Khalifa, a petrochemical plant in Taiwan, or Masdar City. This makes the conversation fictional as well. However, its content is very pertinent to megaproject management in general. I have also taken some liberty to play with cultural stereotypes.

1.2 Status of Megaprojects

Megaprojects are a fact in our daily lives, and they are part of our history. The Egyptians built the Great Pyramid of Giza almost 4700 years ago, and it was definitely a megaproject (Figure 1.3).

Six of the seven wonders of the ancient world were megaprojects at the time of construction (Great Pyramid at Giza, Hanging Gardens of Babylon, Temple of Artemis at Ephesus, Mausoleum at Halicarnassus, Colossus of Rhodes, and Lighthouse of Alexandria). Both admiration and big challenges are associated with ancient and recent megaprojects as the introductory discussion among the managers of Burj Khalifa, the petrochemical plant, and Masdar City illustrates.

Figure 1.3 The Great Pyramid of Giza in Egypt. Source: sculpies/Shutterstock.com.

In most countries around the world, there is at least one megaproject under construction. In many countries, daily newspapers report on several megaprojects on a regular basis. Flyvbjerg et al. (2003, p. 1) wrote, *"Wherever we go in the world, we are confronted with a new political and physical animal: the multibillion-dollar mega infrastructure project."* While I can accept the described widespread occurrence of megaprojects nowadays, I cannot support the label "new." Further historical examples of megaprojects are the Suez or the Panama Canals, the cathedrals of the Middle Ages, the Pont du Gard, or the Colosseum in Rome.

Greiman (2013, p. 11) stresses the growing importance of megaprojects: *"Megaprojects are growing at a fast pace, not only in the United States but in all corners of the world."* The main drivers of the demand for megaprojects are trends toward urbanization and globalization. In 1950, 30% of the world's population resided in urban areas; in 2014, this percentage increased to 54%; and by 2050, it is projected to reach 66%. Together with a growing world population, this means that approximately 7 billion people will live in existing or new cities by 2050 (Figure 1.4). A main facilitator of this is the relative abundance of capital to finance more megaprojects than ever before.

This massive movement from rural to urban areas entails construction demands: new cities or new quarters in cities (living, servicing, manufacturing, entertaining, and worshiping), new infrastructure (roads, rail, and utilities), and new logistical hubs (stations, ports, and airports). Globalization requires connections between hubs and shifts facilities from one country to another (e.g. manufacturing or servicing). There are also truly global megaprojects such as China's New Silk Road (Belt and Road Initiative).

Flyvbjerg et al. (2003, p. 1) characterize megaprojects as "animals." Grün (2004a) speaks of the "taming of the unruly" using a similar terminology. Grün (2004b) clarifies what "taming of the unruly" means by calling megaprojects the "giants among projects" and by concluding "big projects – big problems," which leads to the consequence of giant projects as gigantic problems.

What might look terrifying to academics might not scare practitioners in the same way. However, this is not true for megaprojects. Alex, Choi, and Khaled discussed big problems, chaos, and just keeping their heads above the water. The threat of drowning is definitely frightening!

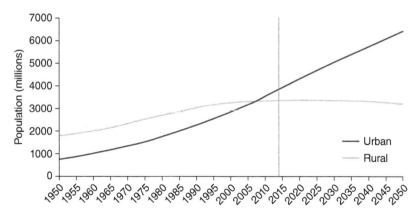

Figure 1.4 Global urbanization trends (United Nations 2014). Source: United Nations (eds.) (2014). World Urbanization Prospects, 2014 Revision, Highlights. New York, United Nations.

Desjardins (2017) provides a list of the seven largest projects under construction in 2017:

1. Al Maktoum International Airport (UAE, 82 billion USD): Once the airport is completed, it will be the largest in the world.
2. South–to–North Water Transfer Project (China, 78 billion USD): Three canals transporting water from the Yangtze River to areas in the north.
3. California High-Speed Rail (USA, 70 billion USD): A high-speed rail link from Sacramento to San Diego, passing through San Francisco and Los Angeles.
4. DubaiLand (UAE, 64 billion USD): Theme park, hotels, and malls.
5. London Crossrail Project (UK, 23 billion USD): Expansion of London's underground transportation system.
6. Beijing Daxing International Airport (China, 13 billion USD): The second international airport in Beijing.
7. Jubail II (Saudi Arabia, 11 billion USD). Jubail Industrial City with industrial plants, an oil refinery, and a desalination plant.

This list is most likely incomplete, since nobody collects information about megaprojects systematically. However, these projects show what civil engineers do: planning and constructing the built environment. The built and the natural environments have a common border. When we expand the built environment, the natural environment will suffer. There are trade-offs to pay while establishing the borderline; it would be naïve to admire the size and grandeur of megaprojects without thinking about their impact on the natural environment.

A case in point is the California High-Speed Rail. It has the positive ecological impact of reducing CO_2 emissions, and it will destroy certain biospheres. It has a political impact as business executives and the Democratic Party support the project in general. Farmers and the Republican Party would rather spend the money on irrigation systems with another set of ecological impacts. Sourcing for megaprojects is global. Tariffs imposed by the Trump administration (e.g. 25% on imported steel) increased the costs for the project considerably. The federal administration has renounced funding on the project. We find that megaprojects trigger ecological, economical, judicial, political, and social impacts.

In sum, megaprojects are part of our history; they have become more prevalent in our times and cause gigantic problems. It is also evident that the "taming of the unruly" continues, which implies that we have not solved the problem of megaproject management. We need to expand and sharpen the tools and approaches of project and construction management for application in megaprojects. This calls for advancing construction project management.

1.3 Purpose

The main problem that managers and engineers are facing is an overwhelming complexity combined with the singularity of a megaproject (Grün 2004b). The singularity of megaprojects implies that previous methods and tools from experience will not provide the best solution for the new project. Using tried and tested approaches will lead to suboptimal results or even outright failure. Very few buildings or civil engineering structures are the same. However, there are degrees of singularity ranging from rather small to very large. This statement must be confusing to a linguist because "singular" describes something that only exists once. The same holds true for "unique." However, degrees of singularity arise when we evaluate different aspects. A bridge might be very long but of limited height. Such a bridge is less singular than one which is a just a bit shorter but much higher. These are just two aspects; however, many aspects are pertinent for the description of megaprojects. Very large singularity characterizes most megaprojects. Civil engineers must choose from billions of parts and combine millions of them; the possible combinations are countless. The "parts" not only relate to the technical aspects of a project but also to social and cultural ones. For each project, we need to develop a special understanding and approach. On a higher level, this book is therefore about complexity management at the limits of human understanding – about taming the unruly.

We also need to analyze separately the configuration of all involved stakeholders with their different values and goals on the one hand and the available technical and managerial solutions for every megaproject on the other hand. An experienced project manager describes the situation at the start of a megaproject in the following way:

Beginning of a megaproject = (quote) "But what happens in the start of these projects is that you suddenly have to throw a medium to large company together with no procedures, no processes, no understanding, and no trust, and you have to throw it into being as an operational organization from day one. And so, you have a situation where nobody really knows what the other person is doing, why they are doing it, how they are doing it, and even if they should be doing it."

Thomas, Australian project manager

Thomas led the project to a successful end. From the quote, we learn about the size of the project by referring to a medium or large company. The project manager and his team needed to realize a turnover of more than 700 million USD within the first year. A volume of 700 million USD per year characterizes a company among the top 10 contractors in most countries around the globe. While these top 10 contractors had many years to grow, the megaproject needs to be "an operational unit from day one" without having an organization. This describes pure chaos. Because of the singularity, nobody knows exactly what to do, and because of the complexity, learning to know what to do is not easy.

I can summarize the purpose of this book in a question: What do we need to do in a megaproject to get from a start where nobody knows what to do to a successful end? This is what I want to explain.

1.4 Methodological Approach

I have worked for more than 20 years as a practitioner in construction and another 20 years as an academic at universities. As practitioner, I was involved in megaprojects in Denmark,

Thailand, Qatar, and Egypt. As an academic, I have researched and taught on megaprojects in Germany, USA, UK, Thailand, Vietnam, and Myanmar. Combining these perspectives, I can only follow Mintzberg (2009) to describe management as a practice. All our knowledge about managing megaprojects is rooted in the knowledge of practitioners.

Thus, I take a descriptive approach. My data derives from direct experience, participant observation, formal and informal interviews, as well as project documents of all kinds. As a project manager for the Bang Na expressway in Thailand, the largest bridge in the world, I managed the project from the beginning almost to the end. We found a way from universal chaos to a rather well-structured performance. What we did not have was time for reflection or for explanation, and thus for organizing and canonizing our experience. Becoming an academic gave me this opportunity. I stopped practicing and started theorizing. What I have learned from this personal experience is that academics are well equipped to structure knowledge from empirical data. What does not work in management writing is to come to conclusions by cognitive deduction; such a normative approach assumes management to be a science, which it is not.

Academic writing aims to be precise based on clear definitions and causal links. Watzlawick et al. (1967) ascribe these attributes to what they term "digital communication." Academics refer to research by others through referencing, thus building theory. Many believe academic language can be objective by refraining from any personal allusions. The personal "I" is then avoided by using the passive voice ("it can be concluded…"), referring to a collective ("we found…"), or by abstracts ("one can deduct…"). As an academic, I feel this rather obscures the source – and, as a practitioner, I dislike this style.

Discussions among practitioners often take the form of storytelling. Stories not only convey content but also emotions and morals, attributes of what Watzlawick et al. call "analog communication." Stories are lacking in precision but have more reach. In this book, I will freely mix the two modes of digital and analog communication. A purist might not like this approach, but I feel that I have no other choice, as I aim to address practitioners as well as academics.

Readers with academic inclinations will find further elaborations on methodology in Chapter 2 (theoretical background). Other readers might want to skip that chapter.

When I read academic books, I am often not sure what precisely the author is talking about. I find that definitions are missing. Therefore, I will introduce definitions formally and highlight them in the text.

 Introducing definitions = (def.) Definitions are indicated in this way. Typically, I will create definitions that mostly apply to this book. My definitions will be nominal – that is, name-giving definitions. Names and nominal definitions are context-sensitive.

Sometimes, I will refer to other authors or people that I have interviewed by quoting them directly. On the one hand, I will do this when authors have phrased their idea in a way that seems perfect to me. On the other hand, I believe that quotes from practitioners provide authenticity and are the raw data for many of my explanations and expressed views. Quotes also have a special format.

 Direct quotations = (quote) "Most of the time indirect quotations are better suited for writing because direct ones are embedded in a different context. However, sometimes a quote is just perfect. Direct quotations must name the source."

Source

Practitioners rely on stories by using analog communication. From an academic point of view, stories provide only anecdotal evidence, that is, evidence that we cannot trust academically. Geertz (1973) introduced the term "thick description" to ethnographic methodology. People in megaprojects form a very specific "tribe," an ethnographic unit. Following Geertz, a description of megaprojects requires the use of stories or observations. I will introduce them formally as follows:

 Observations = (obs.) Observations are stories that provide a thick description. As all stories, they are subjective. However, they might and should be exemplary and give a thick description.

1.5 Readership

In 1995, I was flying business class from Frankfurt to Bangkok during the night. There were more flight attendants than passengers, and we enjoyed all imaginable comforts. I should have slept well. Alas, I did not sleep at all! The next day was to be my first working day as a project manager for a megaproject, my first day with a new construction company, and my first day in Asia. I could not turn my thoughts away from the question of what it means to manage a megaproject in a multicultural environment. Back home in Germany, I had met with engineers and managers with international experience from a number of megaprojects on all continents. They could not answer my questions on how to manage a megaproject. Their knowledge was implicit. At that time, I dearly wished I had a book that explained megaproject construction management in some detail.

Such a book does not exist even today. There are a number of authors researching and describing the owner's perspective with a specific focus on the front-end of megaprojects. These books are of little help to contractors executing megaprojects. This lack is surprising, since there are considerably more engineers and managers working on the contractor's side than on the owner's, and they deal with very different issues. On the Qatar Integrated

Railway Project, approximately 1000 employees work directly for the owner while 5000 work for contractors, managing approximately 50 000 workers.

The following groups of readers will benefit from reading the book in its entirety or in parts.

1.5.1 Managers and Engineers Working for Construction Companies

Managers and engineers working for construction companies form the main audience for whom I write this book; they will find explicit knowledge on megaproject construction management. The bulk of knowledge originates from my experience formed by looking over the shoulders of peers on past projects. At the moment, knowledge about megaproject management is mostly implicit and handed over in projects. Existing codified knowledge, such as that laid out in the Project Management Body of Knowledge (PMBOK 2017), only scratches the surface of what is required for megaproject management. Documenting and discussing the implicit knowledge of project participants allow for the description of advanced construction project management and open it for critical discussion.

1.5.2 Owners of Megaprojects

In order to shape and control megaprojects effectively, owners must understand how contractors spend the largest amount of their investment. The biggest risks materialize during construction, even if we can trace their root causes to the front-end. Planning and execution work together like the two blades of scissors. This book – with its focus on execution – requires knowledge of the front-end as its counterpart, and the opposite is true as well. The success of a project depends on both planning and constructing.

1.5.3 Designers of Megaprojects

Megaprojects are most often civil engineering projects where product and process design are closely interlinked. Design optimization requires the consideration of construction processes to achieve cost efficiency and buildability. The design of the Sydney Opera House caused many problems during its construction; it was very difficult to implement. In my opinion, the breathtaking structure justifies the 16-fold cost explosion. However, this is not true for the majority of megaprojects with problematic designs. Designers must understand the repercussions of their decisions from the end of the project. This is especially true for design/build projects where product planning overlaps with process planning and execution.

1.5.4 Project Managers and Quantity Surveyors Working for the Owner

Project managers (PM) and quantity surveyors (QS) are heavily involved in planning and controlling all processes of megaprojects. They are an integral part of the implementation processes. A deep understanding of the construction phase is indispensable for them.

1.5.5   Managers and Engineers of Large Civil Engineering Projects

Megaprojects act as magnifying glasses for general construction project management. Because of their complexity, few problems remain hidden. Each problem might (and often does) bear heavy consequences, which would not manifest in smaller projects. Lessons learned from megaprojects will provide managers and engineers of smaller projects with the foresight to avoid unaccustomed pitfalls.

1.5.6   Lecturers and Students

Megaprojects are eye-catching and frequently discussed in the press. This generates high interest among students. After completing introductory courses in project or construction management, students can deepen their understanding of these topics while dreaming of their futures. Lecturers will need to prepare for this demand. I have taught courses on megaproject management at Stanford University (USA), University of Reading (UK), in an executive master program of the Asian Institute of Technology in Vietnam and Myanmar, as well as at my own university, the University of Applied Sciences Bremen in Germany. These optional courses always found wide interest among students.

The structure of this book is not like a textbook. However, at the center is a management model, and I will discuss the elements of the model in sequence. This allows for easy adaptation at the university level.

1.5.7   Academe

Academic interest in megaprojects is large while often focusing on policy implications (macro-view). To some researchers, the micro-view of contractors might be of interest as it allows understanding complexity. While the theoretical concepts and the research methodology in this book are academic, some of the language is not scientific (cf. the discussion at the introduction). One practitioner at a premier academic conference summed up his impression about academics by remarking, "*You people talk funny.*" Well, this book on megaprojects intends to give sound and practical help to practitioners. Academics, in turn, might find the walk and talk funny at times. Those who are willing to put up with this funny talk will find new fields for promising research.

Table 1.1 summarizes the value of the book to the different groups of readers.

1.6   Structure of the Text

Chapter 2 (titled 'Theoretical Background') develops the theoretical background that forms the foundation for all the following chapters. As an academic, I need such a foundation to have a clear understanding of my assumptions. Values and a view of the world always generate assumptions. Spelling out the assumptions allows readers to check and compare them with their own views. It also permits a discussion of the validity of different views and

Table 1.1 Value of this book to different readers.

	Practitioners	Academe
Managers/engineers in megaprojects	Core readership	
Owners of megaprojects	Core readership	
PM/QS supporting owners	Highly beneficial	
Designers of megaprojects	Very beneficial	
Managers/engineers of large projects	Beneficial	
Managers/engineers of small projects	Of interest	
International students		Of high interest
Professors and lecturers		Helpful

values. From teaching practitioners in Asia, I know that I hold a Western view of the world that Asians cannot accept in its entirety. However, the data used in this book come from the East as well as the West. Practitioners and students might want to skip this chapter. Academics might want to read it and will certainly find points to disagree with.

Chapter 3 (titled "Advanced Construction Project Management") tries to explain the components of the book title. First, it introduces some basic definitions and the required knowledge to go on reading an advanced book. Next, it looks at construction as an industry with certain idiosyncrasies. This explains my belief that we need not only project management but also, more specifically, construction project management. Moreover, the chapter dwells on how I view management. There are so many books on management with so many different views that it seems indispensable to take a stance.

Chapter 4 (titled "Characteristics of Megaprojects") provides further definitions on projects and complexity to minimize the ambiguity of the text. It also expounds a typology of projects that allows separating megaprojects from others. Finally, it discusses the construct of complexity. This construct is key to understanding and managing megaprojects. Again, academics might find it interesting, and practitioners as well as students less so. If this is true, then a word of warning is necessary: one can manage a megaproject without referring to complexity, but not to its full potential, because the understanding would be lacking.

Chapter 5 (titled "International Construction Management") places megaprojects in the context of the international construction industry and discusses constraints on the management approach imposed by this project-based industry. International or national construction joint ventures usually implement megaprojects, but not exclusively. In fact, very few contractors are strong enough to execute a megaproject single-handedly. This chapter is not essential for academics or practitioners, although both will benefit. For students, this is a must-read.

Chapter 6 (titled "Megaproject Phases and Activity Groups") develops a sequence of phases in megaprojects from first conception to operation. It touches on the activities that precede the signing of a contract. While execution of works starts with the signature, many decisions made earlier impact the execution. These decisions form expectations

Table 1.2 Priorities for different groups of readers.

	1	2	3	4	5	6	7	8	9	10	11	12	13	14
Contractors in megaprojects	For all	+	+++	+++	+++	+++	+++	+++	+++	+++	+++	+++	+++	For all
Contractors in general		+	+++	++	+	++	+++	+++	++	++	++	+	+++	
Owners of megaprojects		+	++	+++	+++	+++	+++	+++	+++	+++	+++	+++	++	
Designer, PM, and QS in megaprojects		+	+	++	+	++	+++	+++	++	+++	++	++	+++	
Students, lecturers		++	++	+++	+++	+++	+++	+++	+++	+++	+++	+++	+++	
Academics		+++	+	+++	+	++	+++	++	++	++	++	+	+++	

and restrictions. It will also become clear that the front-end and execution of works are mutually interdependent. There is no one-sided sequential impact from the front-end to later stages. I believe that this chapter will be beneficial to all groups of readers.

Chapter 7 (titled "Descriptive Megaproject Management Model") introduces the model that forms the backbone of the book. It serves to describe the purpose of the model as well as the importance of all elements and their interactions. It is a synopsis of what follows later. When using this book as a textbook, this and the following chapters are essential. The contents of this chapter will help practitioners manage megaprojects better. The same holds true for Chapters 8–13.

Chapter 8 (titled "Engineering Management") advances the civil engineering knowledge required for megaproject management from the basic courses taught in most universities. The knowledge pertains to design and design management, project management, production planning, site installation, and managing construction.

Chapter 9 (titled "Management Functions") places the often-discussed activities of planning, organizing, staffing, directing, and controlling in the context of megaprojects. These activities form the core of knowledge in business administration. Unfortunately, most managers and engineers in megaprojects have had little exposure to these thoughts. It becomes necessary to repeat and adjust the management functions to the environment of megaprojects. Managers perform at least some of these functions every day in ever-changing sequences and with ever-shifting attention.

Chapter 10 (titled "Meta-Functions") introduces the activities of decision-making, communication, coordination, and learning, which are going on continuously, sometimes in the foreground and sometimes in the background. As Watzlawick et al. (1967) said of communication: *"You cannot not communicate."* The reasons for presenting these functions are the same as for the management functions: managers and engineers have had little exposure to them, and we need to adjust them to the environment of megaprojects.

Chapter 11 (titled "Basic Functions") continues with the explanation of project knowledge, trust, sense-making, and commitment. It stresses on the importance of factors that managers often overlook. Yet, without the right amount and the right mix of basic functions, megaprojects are doomed to fail.

Chapter 12 (titled "Cultural Management") takes a somewhat fresh look at cultural management. Most experienced megaproject managers see cultural diversity not as a problem but as an indispensable component for success. The chapter describes a process that allows maximizing the positive and minimizing the negative aspects of cultural diversity. By analyzing culture, we can also become better at distinguishing between behavior that is culturally determined or a personality trait.

Chapter 13 (titled "Innovation in Construction Megaprojects") focuses on an important dynamic factor. The singularity of megaprojects makes innovation a necessity. In megaprojects, we often use not only cutting-edge technology but also new approaches to management. While I believe this chapter to be beneficial to all readers, I feel it might be of specific interest to academics.

Chapter 14 (titled "All in All, What Does it Mean?") provides a conclusion from a different perspective: What constitutes advanced construction management with regard to megaprojects? What is all the fuss about?

Table 1.2 summarizes the priorities for different groups of readers.

2

Theoretical Background

In this book, I will mix logic and myth, data and stories. Both, logic and myth have their own rationality. While academic professionals prefer basing their conclusions on data, construction professionals often use stories to explain and provide leadership. Data are more reliable and able to cut through a thicket of obscure networks of factual relationships to expose causality; however, anyone who understands data from a complex social environment as objective is plainly naïve. Data are also sometimes limited in reach. Stories are rich in content, albeit sometimes purely subjective, i.e. meaningful to only one (deranged?) person. The pervasiveness of stories in the construction industry – small or large – must impress anyone who has encountered it. To disregard stories means to refuse understanding. A bridge between data and stories are "thick descriptions" (Geertz 1973). Thick descriptions filter contextual data (versus raw data) from stories.

Two philosophical ontologies are hidden behind the juxtaposition of data and stories: the objective external world governed by natural laws (Bacon 1620) and the socially constructed world (Berkeley 1710, *esse est percipi*). Instead of taking one of these exclusive positions, I prefer to follow Popper's Three World Theory (Popper and Eccles 1977). According to this view, there exists a World 1 of physical objects (external world), a World 2 of psychological states (internal world with thoughts, emotions, and impressions), and World 3 (intersubjective knowledge composed of theories, artifacts, myths, and religions). These three worlds are interrelated and influence each other in different ways. I believe that we must concentrate on World 3, especially when researching management, as it contains both relevant data and stories.

2.1 Definitions

We learn our everyday language by trial and error during the different stages of our socialization. In schools and universities, we learn a more professional use of specific terms. However, these terms often lack codification or acceptance of any existing codification. This might lead to misunderstandings, especially when different professional languages mix. In construction projects, this could be business and engineering language. Accordingly, we need definitions to formulate our thoughts more precisely.

Advanced Construction Project Management: The Complexity of Megaprojects,
First Edition. Christian Brockmann.
© 2021 John Wiley & Sons Ltd. Published 2021 by John Wiley & Sons Ltd.

In philosophy, we distinguish between nominal and real definitions. The latter describe the essential attributes of a term (definiendum). This is quite a challenging task and, in my opinion, it would be better to leave this task to philosophers. Real definitions are either true or false, and this makes them difficult to formulate. Nominal definitions are much easier to express; they are either useful or not useful in the sense of the user. They are also practical as they replace a long explanation (definiens) with a single term (definiendum) or a name. To avoid confusion, the defined term should be similar to everyday language. As nominal definitions need to be useful to the person developing them, this person has some freedom when formulating a definition.

Nominal definition = (def.) In modern science, only nominalist definitions occur, that is to say, shorthand symbols or labels are introduced to cut a long story short.

Popper (1945, 2002, p. 16)

Thus, we find a long story as the right-hand part of the definition and a shorthand label to the left. I will formally emphasize definitions in the following way: Term = (def.) long explanation.

2.2 Cognitive Maps

Maps are always a reduction of reality, giving only relevant information and omitting much of the accessible yet unnecessary information. Geographical maps are dynamic; they change as we build new roads, railways, canals, new quarters, or cities. Less frequently, they require corrections for the course of a river, the eroded elevation of a mountain, or the waterline due to a rise of the sea level. In sum, maps are a simplified and dynamic representation of reality.

Cognitive maps store information in our brain. This information is simplified and needs adjustments from time to time (Tolman 1948). Other words for cognitive maps are scripts or frames of reference. I prefer the term "cognitive map" because of the characteristics of maps: dynamic representations of the real world.

Cognitive map = (def.) A cognitive map is a mental understanding of an environment, formed through trial and error as well as observation. The concept is based on the assumption that an individual seeks and collects contextual clues, such as environmental relationships, rather than acting as a passive receptor of information needed to achieve a goal. Human beings and other animals have well-developed cognitive maps that contain spatial information enabling them to orient themselves and find their way in the real world; symbolism and meaning are also contained in such maps.

American Psychological Association (in van den Bos 2007, *p. 190)*

Nussbaum (2018) adds that cognitive maps also contain ethical information on such categories as good or bad.

Managers of megaprojects assemble cognitive maps through observation as well as trial and error. Because of learning processes, such maps of experienced managers contain the best knowledge on how to solve problems encountered in construction megaprojects. This knowledge enables them to navigate through the maze of uncertainties, ambiguities, and complexities of megaprojects.

Thus, cognitive maps are the information treasure troves that I set out to find. Aligning the cognitive maps of project participants to those of the most successful megaproject managers provides us with a very powerful tool to organize a coordinated approach toward the goals.

2.3 Descriptive Management Research

Descriptive management research collects data from managers and is interested in their thoughts and actions. It assumes that managers create best practices and an understanding by solving management problems daily. The task of the researcher is to observe, record, conceptualize, and transform implicit knowledge into explicit knowledge. The opposite is normative management research, where the researcher sets norms on how to perform management. I strongly believe normative research in management to be a wrong approach because such researchers live in a theoretical world without any grounding in practice.

I want to understand what construction managers think and do in megaprojects. I want to discover their cognitive maps. Typically, these managers cannot give clear descriptions of their cognitive maps as they form part of their implicit knowledge. The unrelenting pace of management does not allow them to reflect on their cognitive maps. It takes all their available energies to build and correct them through trial and error as well as observation.

My personal advantage is that I have been an integral part of megaprojects day in, day out for many years. I have built my own cognitive maps and adjusted them in discussions with others. There is certainly no lack of exposure, and I doubt whether any other researcher has spent more time in megaprojects with more access to different levels of the hierarchy and privy documents. This involvement, however, carries the risk of bias.

To balance the risk, I have conducted 35 peer interviews, mostly with peers from the Taiwan High Speed Railway Project (THSR). Ethnographic interviews help us to understand a different world and assume that the worldview and the language between interviewee and interviewer differ, i.e. a sociologist interviews a construction manager (Spradley 1979). This is not my problem; as a project manager, I have interviewed other project managers. The interviews were semi-structured, allowing interviewees to recount stories and provide facts and inside perspectives. When I challenged them on their views, they saw me as a colleague.

I have taped the interviews and transcribed them. For evaluation, I used grounded theory (Strauss and Corbin 1998). Grounded theory starts at thick descriptions (Geertz 1973) and uses various forms of coding to extract and structure the contents of the thick descriptions. For the more direct interview questions, I used qualitative data analysis (Miles and Huberman 1994) or simple statistical tools.

I can illuminate the advantages of this process with a story. While interviewing the project director William at the THSR, we somehow stuck to discussing trust – a perpetual favorite of academic literature. William was no academic and asked me why we do not proceed to something more important such as communication. Shamefully, I have to admit that, at that time, communication was not on my list nor in my model, although I perfectly understand and understood its importance – a mental slip that William helped me correct. Thanks to him, you will now find communication to be a part of my model. This story taught me to look humbly at my experiences and listen carefully to others.

2.4 Guiding Theories

Theories are very practical. Different theories allow for different perspectives, and I will profit from this fact. Three theories are essential: Luhmannian systems theory, contingency theory, and new institutional economics (NIE), which has another three sub-theories (property right theory, transaction cost theory, and principal/agent theory).

2.4.1 Luhmannian Systems Theory

Parsons (1991) tried to develop a general theory of society. Today, we call this the structural-functional systems theory; accordingly, structures characterize systems. Structures are rigid, and this earned the theory a negative reception, as systems are flexible. Luhmann as his student improved the theory and changed it to the functional-structural systems theory. What looks like a play of words has profound implications. On the one hand, few scholars still refer to Parsons. On the other hand, Luhmannian systems theory (2013) has become one of the most influential sociological theories in Europe. The success is due to the interest in the functioning of systems and the inclusion of the environment. Luhmann started researching open systems before switching to operatively closed systems; such systems develop autopoietically, i.e. in referring only to themselves without connection to the environment. When we conceptualize a construction project as a system, it would be absurd to refer to operational closeness or autopoiesis. Instead, a concept of an open system with interaction between the construction project and its environment is more helpful. Owners develop construction projects and, to a large degree, determine the borders of the system by differentiating them from the environment. Borders are dynamic and require continuous adaptation; this includes scope, cost, schedule quality, and safety. They need to be adapted due to learning (changes of understanding) and altering circumstances (changes of facts). Risk compensation accompanies differentiation to create the system and adaptation to align it to the environment. The determination of borders, specifically for megaprojects, is very risky and needs close controls. By differentiating the system (construction project) from the environment, we can reduce system complexity, albeit with the consequence of increasing project risks. There is a need to manage the remaining project complexity (eigencomplexity), and this is a heck of a task for megaprojects! In sum, Luhmannian systems theory explains the creation of projects with the explicit goal to reduce environmental complexity, and it rests on four columns: (i) differentiation, (ii) management of eigencomplexity, (iii) risk compensation, and (iv) adaptation (Figure 2.1).

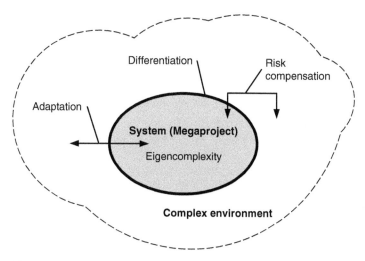

Figure 2.1 Luhmannian systems theory and megaprojects.

In construction projects, contractors create systems as well as owners. Make-or-buy decisions are an example for contractors. Relying on oneself for execution or managing entirely subcontracted works (both extremes) create very different systems with a different balance of management, risk allocation, and learning.

Some researchers use the term "complexity theory" instead of systems theory, highlighting the central importance of complexity management. Systems theory models the environment as complex, contingent, and full of conflicting demands.

2.4.2 Contingency Theory

The term contingency theory connects systems and the environment. Contingency describes a world where alternative actions are possible. This presumes freedom of human action and available alternatives. In systems theory, this signifies freedom to alternative differentiations between the system and the environment. When managing eigencomplexity, it means freedom to create different subsystems. Contingency excludes the impossible and the necessary. It describes the relationship between the environment and the system as well as the system and subsystems. The possibility of choice does not entail arbitrariness; limiting boundary conditions are very real.

Complexity and contingency allow for two different forms of conflicts. On the one hand, differentiation between the environment and the system creates potential for external conflicts, and on the other hand, contingent action allows for internal conflicts when reducing the eigencomplexity of the system.

Contingency theory tries to determine what frames our possibilities of action and structuring. The central idea is one of situational influences. Contingency theory pays special attention to organizational matters. According to the theory, formal organizational structures have a strong influence on efficiency. As there is no universally efficient structure, these must always align with situational factors.

Important results of the contingency theory stem from research by Woodward (1965) on the influence of production technology and Fiedler (1964) on the influence of leadership. Criticism of the contingency theory points to five problems: (i) there are examples of successful organization that are not situationally determined, (ii) the theory does not explain adaptation processes, (iii) effects of power do not play any role, (iv) it entails conservative structures, and (v) there are no objective descriptions of situational factors independent of actors.

Mintzberg (2009) has advanced contingency theory by including actors. This way, he derives from theory six organizational types: (1) entrepreneurial organization, (2) machine organization, (3) professional organization, (4) project organization, (5) missionary organization, and (6) political organization. The choice of an organizational type depends on situational factors. Accordingly, the relationship between situational factors and efficiency stops being deterministic. Stuart (2002) sees increased interest in approaches using contingency theory embedded in systems theory of open systems.

There are answers to the preceding criticisms. Uncoupling the relationship between situational factors and efficiency makes different outcomes possible and plausible. The inclusion of actors allows for adaptation and subjective descriptions of the situation. However, neither systems theory nor contingency theory can explain the use of power. Principal/agent theory allows doing just that, and a description of this theory will follow in the next chapter.

I will make use of contingency theory embedded in systems theory in a way where contingency means both dependency and limited possibility. The four columns of systems theory (differentiation, eigencomplexity, risk compensation, and adaptation) frame the following tasks in megaprojects:

- Differentiation: project development, scope management, and stakeholder management
- Eigencomplexity: decentralization and management
- Risk compensation: controlling and risk management
- Adaptation: learning

2.4.3 New Institutional Economics

Institutions shape construction projects in many ways. The simplest definition is that institutions determine the rules of the game (North 1990). Schotter (1981) introduced a formal definition:

 Institution = (def.) "A regularity R in the behavior of members of a population P when they are agents in a recurrent situation Γ is an institution if and only if it is true that it is common knowledge in P that

1. everyone conforms to R;
2. everyone expects everyone else to conform to R; and
3. either everyone prefers to conform to R on the condition that the others do, if Γ is a coordination problem, in which case uniform conformity to R is a coordination equilibrium; or
3. if anyone ever deviates from R, it is known that some or all of the others will also deviate and the payoffs associated with the recurrent play of Γ

using these deviating strategies are worse for all agents than the payoff associated with R."

Schotter (1981, p. 11)

This definition clarifies that game theory can be a preferred means for the analysis of institutions. There are five different types of rules (or regularities). First, there are conventions observed by self-control. Second, we have ethical rules with which we comply by imperative obligation; Kant's categorical imperative is an example with regard to generally applicable laws. Third, others impose the observation of customs. Fourth, control by others enforces following private formal rules; organizational rules belong to this group. Fifth, law enforcement agencies demand the observance of public laws. The degree of observance of rules or institutions can explain differences in welfare between countries (Acemoglu and Robinson 2012).

Neoclassical economics work with a model of man called *homo economicus*. This actor knows the future and, because of this knowledge, can choose the best alternative without spending time or resources. Few economists see this as a realistic description, but models like supply and demand provide reasonable results based on the behavior of a *homo economicus*.

The NIE assume different characteristics for economic actors. They do not know the future, and accordingly, have to rely on incomplete information. In addition, limited rationality describes their behavior and decision-making. They try to act rationally but sometimes fail. Maximizing benefits becomes impossible; instead, the actors of NIE employ satisficing, the choice of an acceptable solution (Simon 1955). Furthermore, market exchanges are no longer without price; there are transaction costs (Williamson 1985). Anyone who has been an economic actor will feel much more comfortable with the assumptions of NIE. Behavioral economics nudge the model of man in an even more realistic direction by describing the ways in which we fail to behave rationally (Thaler 2015).

The modeling of actors in NIE led to three different theoretical approaches. The inclusion of external effects, such as the free use of water from a river to cool a plant, that played no role in neoclassical thinking led to the property rights theory; the negligence of market costs to the transaction cost theory; and incomplete information to the principal/agent theory.

The theory of property rights assumes profit- or utility-maximizing behavior and transaction costs for writing and enforcing contracts. Connected to property rights is the theory of incomplete contracts (Grossman and Hart 1986). Picot et al. (2015) cite construction contracts as prime examples of incomplete contracts. The main conclusion is that incomplete consideration of property rights leads to external effects and, consequently, to inefficient factor allocation.

Risk allocation in construction contracts is a constant concern, and the property rights theory demands that the allocation of risks is complete. As risks are events in the future and nobody can foresee the future, there will be risks that are not part of the contract. According to the theory, this will lead to inefficiencies.

The theory of transaction costs adds transaction costs to production costs. Wallis and North (1986) researched the amount of transaction costs of private enterprises as a percentage of GDP. They found an increase from 20% in 1870 to 40% in 1970; transaction costs are substantial and increase with economic development. Efficiency must consider both production and transaction costs as they depend on different institutional arrangements. Actors are limitedly rational, opportunistic, and risk-averse. In consequence, efficient institutions minimize transaction costs and, at the same time, safeguard against opportunistic behavior. Too much supervision in construction projects costs too much.

It is difficult to determine transaction costs exactly. If transaction costs include all items connected to writing and enforcing the contract (the owner's responsibilities), production costs encompass design and construction. Greiman (2013, p. 239) provides data for the Boston Artery, where production costs amount to 10 713 billion USD, and the owner's transactions costs amount to 4085 billion USD, i.e. 28% of the total expenditure. Assuming that the designers and contractors face a similar amount of transaction costs during their activities, we arrive at another 2957 billion USD. In total, the net production costs would be 7756 billion USD and the transaction costs would stand at 7042 billion USD – almost 48%. Whatever the inaccuracies in this calculation, transaction costs have high relevance.

The focus of the principal/agency theory are contracts that regulate exchanges. These can be exchanges between organizations (owner and contractor) or within an organization (owner and manager or any other hierarchical relationship). The exchange initiates coordination and motivational problems, which we can solve through incentives, controls, and information. The actors are utility maximizing, limitedly rational with asymmetrical information; they have different risk appetites and tend to display opportunistic behavior. Who would not think of the antagonisms in construction projects between owner and contractor, designer or project manager, or contractors and suppliers or subcontractors?

Flyvbjerg et al. (2003) describe typical principal/agent situations when relating the behavior of public owners. In this case, the taxpayer is the principal and the public owner is the agent. They promote changes to accountability to control the problem of cost overruns. Greiman reports a case of flagrant opportunism when the owner of the Boston Artery, the Massachusetts Highway Department (MHD) gave an order to his project management consultant Bechtel/Parsons Brinkerhoff (B/PB) to reduce the cost forecast. The Office of the Inspector General (OIG) revealed the opportunistic behavior of MHD in a report in 2001:

 Opportunism = (quote): "Importantly, the OIG's investigation also revealed that shortly after they had been provided with this up-to-date cost information, in early 1995, the MHD directed its management consultants (B/PB) to reduce projected costs from $13.78 billion to $7.7 billion. Obviously, this could only be done on paper, not in the reality of construction."
Greiman (2013, p. 225)

As it turned out at the end of the project, the sum of 13.78 billion USD was the amount due. Reducing the projected amount helped the owner at some point in time and, for this reason, he chose to ignore reality.

3

Advanced Construction Project Management

The term "construction management" can pertain to three different areas of application and content:

- Management in the construction industry, i.e. management of construction companies
- Management of construction projects
- Management of construction activities

I will use "corporate management" when addressing the company level, "construction project management" when referring to construction projects, and "construction management" when talking about on-site construction activities. This leads to two formal definitions.

 Construction project management = (def.) All required activities that allow achieving the goals of a construction project. This includes initiating, planning, executing, monitoring, and controlling, as well as finishing the project. Depending on the contract type, other activities might become necessary, e.g. design (design/build contracts), financing (design/build/finance contracts), or operations (design/build/operate contracts).

 Construction management = (def.) All activities to confirm that the project execution achieves the stipulated quality, stays within budget, and is delivered on time while ensuring the protection of safety, health, and environment, and also fulfills other stakeholder expectations if possible. This includes labor, subcontractor, and equipment management, and excludes planning of production processes and site installation.

Advanced construction project management enlarges the knowledge of typical courses in construction management or the experience of managers on mid-size projects. The reader

Advanced Construction Project Management: The Complexity of Megaprojects,
First Edition. Christian Brockmann.
© 2021 John Wiley & Sons Ltd. Published 2021 by John Wiley & Sons Ltd.

should possess such knowledge before advancing any further. Harris and McCaffer (2006) cite, among others:

- scoping/budgeting the project
- design coordination/management
- establishing the management structure of the management team
- marketing/procurement
- defining roles/responsibilities
- estimating/tendering
- stakeholder management
- project and construction method planning/coordination/control
- value/risk management
- organizing/leading/implementing controls
- production/productivity management
- management of labor resources/temporary works provisions/equipment/plant/subcontractors/suppliers
- time and subcontractor interface management
- cost and budgetary control, including cash flow forecasting
- quality management
- contract and progress payment administration
- legal issues
- information and communication technology (ICT) management
- health and safety management
- environmental impact management

This is a long and demanding list to start with, but in many countries, these issues are part of bachelor's programs in civil engineering or construction project management. Advancement beyond the listed topics will mostly proceed in the direction of management activities in the context of highly complex construction projects.

3.1 Construction

In certain aspects, the construction industry is similar to other industries, while it differs in others. The basic difference is that we can characterize construction projects as contract goods as opposed to exchange goods, such as cars. Everyone can examine an exchange good before purchase; contractors, on the other hand, produce their (construction) contract goods after agreeing with the owner on a contract. Contract goods have ex-ante no search qualities; we cannot check them at the time of purchase. Additionally, owing to singularity, megaprojects do not provide for experience qualities; owners seldom purchase a similar project from the same contractor. Without the possibility of checking construction goods physically before purchase and without prior experience, the owner must rely on trust qualities. The situation is akin to the opening of a Pandora's Box – full of undesirable phenomena. According to the principal/agent-theory (Jensen and Meckling 1976), these include adverse selection, hidden attributes, hidden action, hidden intentions, hold-up, and moral hazard. Table 3.1 provides more differences.

Table 3.1 Characteristics of contract (construction) vs. exchange goods.

	Characteristics of construction contract goods	Characteristics of exchange goods
Corporate management	Separation of corporate seat and construction sites	Production at corporate seat
Human resource management	On-site assembly and inclement weather	Production in a factory
Financing/investment	Small equity ratio	Larger equity ratio
Accounting	Construction accounts framework and evaluation of half-finished products	Industry accounts framework
Procurement	Small batches, short-term contracts, and decentralized supply	Large batches, long-term contracts, and centralized supply
Marketing	Before production	After production
Organization	Project organization	Process organization
Production	On-demand, single unit construction, and on-site	On-stock, mass fabrication, and in factories

These differences are too important to neglect. Mintzberg postulates three characteristics for management:

Professional managers = (quote) "…managers are not effective; matches are effective … there are no effective managers in general, which also means there is no such thing as a professional manager."

Mintzberg (2009, p. 222)

This quote postulates that there are no "born leaders" who can manage everything by their great personality. Instead, we must understand leadership as a relationship or match. Matches concern tasks, people, and information. The problems with implementing a megaproject often exceed the limits of human understanding. How can anyone without pertinent experience hope to solve them? People working in megaprojects are not only civil engineers; there are also mechanical, electrical, and computer science engineers involved, along with politicians, bureaucrats, businesspersons, bankers, insurers, lawyers, police, and the concerned and interested public. However, at its core, construction megaprojects pose civil engineering problems. To decipher the bulk of information, engineering expertise is necessary. The closer management is to production, the more specific management has to be (Mintzberg 2009). Construction megaprojects are very close to production; what they require are civil engineering managers. Although engineers sometimes struggle with the political or social implications of their projects, politicans never succeed solving engineering problems.

Civil engineering expertise is the base upon which to implement a construction megaproject successfully. The complexity of megaprojects also requires a sound knowledge of the prerequisite management understanding. This is what I will develop over the following chapters. The idea of professional managers is that they can manage everything, from baby food to construction megaprojects. I cannot follow this logic or find equivalent experience: I have been involved in a number of megaprojects, visited some, and read about others, but not once has a professional manager represented the contractor.

3.2 Management

There are few subjects where pertaining literature is more abundant than for management. There are the do-it-yourself books – the ones with simple solutions and the profound ones based on experience, observation, and research. Simple solutions are attractive because they are plausible to large degrees. Most people can follow the line of thought. However, they fall short of providing clarity – and, in case of megaprojects, they fall perilously short.

Books with simple solutions for management offer a few factors and postulate that these will inevitably lead to success. The metaphor used to describe the approach is that of a machine: put in a coin, pull a lever, and walk away with a chocolate bar. To elaborate, the existence of natural laws is presumed in management. Academic success factor research is also widespread. However, these efforts have still not produced unequivocal results except for very general ones (Kieser and Nicolai 2005). Management belongs to Popper's World 3 of intersubjective knowledge and not to the World 1 of physical objects.

My overall observation of megaprojects managers leads to the same conclusions that Mintzberg describes as fact:

Work of managers = (quote) "Study after study has shown that (a) managers work at an unrelenting pace; (b) their activities are typically characterized by brevity, variety, fragmentation, and discontinuity; and (c) they are strongly oriented at action."

Mintzberg (2009, p. 222)

When managers have to solve a variety of non-sequential problems in a short period by initiating action day in and day out, a solution to the management problem can be neither mechanistic nor simple. Megaprojects add complexity, magnitude, and urgency to the problems: there is an awful lot of things to accomplish in a short time.

Management in megaprojects = (obs.) While working in my office, managers and engineers would line up to come in with all kinds of questions. While discussing with one employee or a group, I could see the next waiting their turn. Most of them were very qualified and came with questions on difficult problems. The scope of the questions ranged from design over construction to administration and dealing with suppliers. Concrete recipes, subcontracts, formwork design, equipment purchasing, IT, accounting, financing, car allocation, apartment furnishings, motivation and payment, organizing traditional local ceremonies, and visits by politicians – all needed consideration and answers. The employees wanted information, contacts, and decisions. This happened every day.

Mintzberg (2009) also proposes a general model for management (Figure 3.1). This provides a structured idea on what it means to manage. It does not describe megaprojects specifically. I will introduce a descriptive model for construction megaproject management in Chapter 7. This discussion will include comments on the connection between the two models, as they observe different levels of management activities. Mintzberg's model is quite abstract. I am taking the liberty of adjusting the explanation of the terms to the requirements of megaprojects, while retaining the level of abstraction.

Managing means dealing with information and people as well as acting. According to the model, dealing with information consists of framing, communicating, scheduling, and controlling. Dealing with people means to link and to lead. Finally, acting consists of doing and dealing. Managers need to do all this internally and externally, in the construction project and in collaboration with owners and other stakeholders.

Framing provides the context in which members of megaprojects work. Managers create the framework through decisions, and it pertains to the organization of aspects such as charts, rules, and encouraged behavior, to staffing and other plans. All this guides and contains further actions. It also includes sensemaking as a form of guided understanding and information-giving.

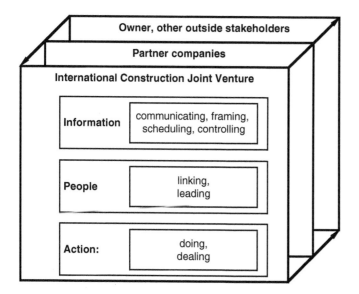

Figure 3.1 General management model.

Managers spend most of their time communicating, sometimes talking, often listening. Many prefer face-to-face communication when receiving or passing on information. Communication increases with project complexity, globalization, electronic media, and virtual teams (Slevin and Pinto 2004). Characteristics of megaprojects include very high complexity; globalized procurement of services, supplies, and equipment with virtual teams spanning the globe; and extensive use of the Internet, email, and video conferencing.

Scheduling not only creates processes that determine action but also allocates resources. This can be the manager's time for a specific person or a problem. It can also describe the labor, material, and equipment allocated to a construction activity. Scheduling organizes action in time.

Controlling is the feedback loop for information given and decisions made. It is not a predominant role for a manager to exercise authority. Information and decisions break down systems into parts. Due to the complexity of megaprojects, there is no way to create a complete set of information at the beginning that fully describes the project. Information develops with time; it is divided by nature. When information comes in bits and pieces, this must also be true for decisions. Controlling allows a synthesis of separate actions based on piecemeal information and decisions toward project goals. Regular schedule controls permit us, for example, to determine whether the completion date has changed or not.

Leadership is a favorite topic of management literature. Much of this literature is American and influenced by American culture, which tends to see leaders as heroes. There are not enough heroes in the world for all the current megaprojects. Accordingly, some megaprojects do without them, and many of those led by heroes, real or perceived, are suffering from their heroics. Leadership is a phenomenon that different cultures interpret differently: "*Leadership is culturally contingent*" (House et al. 2004, p. 7). It seems sensible to understand leadership in global megaprojects as a social situation that can differ from society to society. Megaprojects, with their complexity and urgency, require managers who can delegate with the accompanying loss of authority and control.

Delegation entails linking more than providing direction, i.e. linking people with relevant information and expertise. The manager quoted in Section 1.3 cannot give directions at the start of his megaproject, since he does not know what to do. Finding this out requires humility and a strong ability and motivation to learn. Linking is also required outside hierarchies, i.e. outside the megaproject.

Doing, for managers, does not mean that they go out to cast concrete. Managers are at least one level removed from actually doing site work; project managers in megaprojects are sometimes removed from site work by several levels. Doing, for managers, means going out and looking at a problem directly, getting people together for meetings (not just calling a meeting), and reading up on a problem. It has much to do with problem-solving close to action.

Managers need to deal inside and outside of a megaproject. Dealing refers to the process of balancing the contradicting demands from within and outside the project. Typically, megaprojects are organized as functional organizations. The functional managers pay attention to specific aspects, and, often enough, compromises are the best overall solution. Outside the megaproject, the project manager needs to build coalitions within and between partner companies and sometimes with outside stakeholders. He must also continuously negotiate with the owner (Figure 3.1).

In my view, good managers in construction megaprojects are hard-working and innovative people with a strong inclination to learn and very good communication skills. They might not have solutions, but they will find them. Power is less important to them than problem-solving. They are not heroes and do not see themselves as heroes; they are humble and know that only teams can complete megaprojects successfully.

This general view might be at odds with other management literature, but it is supported by descriptive management research (such as that by Mintzberg).

 Project manager as self-proclaimed hero = (obs.): Some 40 years ago, at a construction site in Aceh, Indonesia, it became necessary to build a camp, since the site was far from any large city. There was not enough housing available. The camp was right on the beach, hidden among palm trees, and the project manager decided the arrangement for the housing. His own house was chosen on the beach, the second level in the hierarchy found their houses one line removed, and so on. The project manager had the organization chart built on the beach – of course, this indicates his desire for power, emphasizing that the project manager is the only hero and that he deserves the benefits belonging to that status.

4

Characteristics of Megaprojects

Various authors attribute different characteristics to megaprojects. Hassan et al. (1999) describe large-scale engineering projects as having the following features:

- high capital costs
- long duration with urgency
- technically and logistically demanding
- multidisciplinary inputs
- a virtual enterprise

Bruzelius et al. (2002) point to high investments of more than 1 billion USD, long lifetime of 50 years and more, considerable uncertainty with respect to demand forecasts and cost estimates, and a considerable share of indirect benefits.

Grün (2004b) stresses on singularity, complexity, goals, and project owners. Accordingly, megaprojects are highly innovative, highly complex, have demanding goals regarding cost and time, and the stakes are high for project owners.

Greiman (2013) provides a rather long list:

- long duration
- large scale
- type of industry
- design and construction complexity
- complicated sponsorship and financing
- ambiguous life cycle
- critical front-end
- high public profile
- public scrutiny
- large-scale policy-making
- innovative contracting and procurement
- discontinuity of management
- technical and procedural complexity
- unique organizational structure
- high degree of regulation
- multiple stakeholders
- dynamic governance structures

Advanced Construction Project Management: The Complexity of Megaprojects,
First Edition. Christian Brockmann.
© 2021 John Wiley & Sons Ltd. Published 2021 by John Wiley & Sons Ltd.

- ethical dilemmas
- consistent cost underestimation and poor performance
- high risks
- socioeconomic impacts
- cultural dimension
- complex systems and methodology
- strong environmental impacts
- collaborative project environment

While the list is extensive, it is also a bit confusing. For example, not all megaprojects have cost overruns. This seems to be more often the case in public projects. Overall, the list reflects, most of all, the lessons learned from the Boston Artery, the Big Dig.

These examples span a large range to define megaprojects. Many authors use an investment threshold to define megaprojects – e.g. Bruzelius et al. (2002) consider this to be 1 billion USD. We can verify such a number easily, and it makes the identification of megaprojects easy. Unfortunately, it is dead wrong. Megaprojects are a global phenomenon; construction costs are local. Workers in Australia cost 60 USD per hour, and in Vietnam less than 1 USD. This decreases the costs dramatically and influences the complexity, since workers in Australia are more productive as they have more and better equipment available to support them.

There is no easy way to describe megaprojects. To solve the conundrum, I will first introduce a theory-based typology of construction projects, allowing a comparison between megaprojects and other projects. Second, I will discuss complexity in the context of construction megaprojects.

4.1 Project Typology

The goal for developing a new typology is to provide a framework that allows us to explain and understand phenomena in the construction industry. Of interest are, of course, the characteristics of megaprojects and their differences from other projects. Trying to create a systematic order of construction projects is of a conceptual nature and means advancing a theoretical contribution. Systematic orders are fundamental elements in the development of a scientific body of knowledge (McKelvey 1975). The lack of a theoretical background of existing orders provides the impetus. A simple example of an existing categorization is the well-known differentiation between residential, industrial, building, and civil engineering projects (Barrie and Paulson 1992; see also Langford and Male 2001). Groups of building projects comprise offices, churches, sports arenas, opera houses, and universities. What do they have in common? Theoretical contributions need to identify building blocks (what), relations (how), and underlying reasoning (why). This reasoning is of special importance because logic replaces data as the basis for an evaluation when building a systematic order (Whetten 1989).

We commonly use four forms of systematic orders: nomenclature, classification, taxonomy, and typology. Nomenclatures in science describe systems of designations. An example

Table 4.1 Characteristics of systematic orders.

	Nomenclature	Classification	Taxonomy	Typology
Ontological base	Definitions	Rationality	Natural laws	Rationality
Assignment by	Definitions	Decision rules	Decision rules	Archetypes
Truth values	No	Yes	Yes	Yes
Empirical content	No	Yes	Yes	Yes
Precise	Yes	Yes	Yes	Yes
Complete	Yes	Yes	Yes	Yes
Disjunctive	No	Yes	Yes	No

is NACE (French: *Nomenclature statistique des activités économiques dans la Communauté Européenne*), the statistical framework used in the EU. Nomenclature as a system of designation is nothing more than a series of nominal definitions. Theory is not the basis of nomenclatures; they have neither truth-values nor empirical content, but they can be practical.

We create classification systems by assigning objects or phenomena to specific classes according to predetermined criteria (similarity or relationships) and decision rules; a hierarchy (Doty and Glick 1994) characterizes them, and they are precise, complete, and disjunctive (not connected).

If natural laws provide the basis for a classification, it becomes a taxonomy. The most widely known classification is the systematic order of plants (Linnaean taxonomy).

Archetypes are the components of typologies. Categories of typologies are precise (however, the assignment is imprecise) and complete but not clearly disjunctive. Unambiguous decision rules for the assignment are not possible. We can distinguish different characteristics for these four systematic orders (Table 4.1).

Clearly, either a classification or a taxonomy would be preferable as they are most stringent. Unfortunately, construction projects do not follow the order of natural laws: establishing a taxonomy is indefensible. A classification, on the other hand, poses problems when the population that it tries to capture is continuous. Project size measured in monetary units would result in distinguishing two construction projects that differ by the amount of one dollar: establishing a classification is undesirable. Some fuzziness about the categories will prove beneficial.

4.1.1 Conceptualizing Criteria

The first step in building a typology is conceptualization. We can achieve this by empirical induction or rational deduction. Induction is a process starting at a concrete level, then becoming more abstract. Theories are the basis for deduction, allowing us to derive dimensions. In applications, we can verify the usefulness of the chosen dimensions. This is a process that is becoming more concrete.

In construction, it is fundamental to differentiate between product and process. The client typically designs the product (building and structure) with the help of architects and engineers, while the contractor develops the construction process. Product and process planning are interdependent; both are part of construction projects.

4.1.2 Choice of Dimensions

The starting point for the selection of possible dimensions for the construction industry are publications analyzing organizations and their tasks. In this business context, authors distinguish between task and resource dimensions. Tasks can be difficult, variable, interdependent, complex and novel or easy, routine, independent, simple, and repetitive. Resources can be specific or generic. I will discuss each of these dimensions.

Difficulty: It is not easy to analyze a difficult task. Algorithms for solutions are unknown (Van de Ven and Delbecq 1974). Perrow (1967) describes difficulty as the degree of complexity of the search process in performing the task, the amount of time required for thinking, and the body of knowledge that provides guidelines for performing the tasks. The underlying theory is task contingency – a special form of contingency theory – i.e. the idea that tasks determine the structure of an organization (Fiedler 1964). This also holds true for the next two dimensions, variability and interdependency.

Variability: Tasks with great variability demand the use of a multitude of different approaches (Perrow 1967).

Interdependency: We can differentiate between interdependency within and between teams. It is high if many members of a team or many teams need to interact to find the solution to a problem. High interdependency increases the demands for coordination (Tushman 1979).

Complexity: Complex tasks can be broken down into many parts, and there are many connections possible between the parts. A third component of complexity is the cross-impact of decisions, the question of consequences for the system. When many decisions entail important consequences, we face high complexity. Multiple parts, interactions, and consequences create uncertainty when solving a problem (Tushman and Nadler 1978). Complexity belongs to a theory that interprets organizations mainly as information processing and decision-making entities. It also plays an important role in systems theory (Luhmann 2013).

Novelty: Novel tasks are the opposite of repetitive tasks; the degree of newness to the actor defines the dimension (Puddicombe 2011). For their solution, we first need to develop new structures, followed by new algorithms. Repetitive tasks are already pre-structured. The construct of novelty also belongs to the realm of contingency theory.

Resource specificity: Of high importance are human, asset, location, and temporal specificity. High resource specificity describes high quality demands in these areas (Williamson 1991). Using the concept of opportunity cost, we can state that resources are very specific if the difference to the second-best alternative use of the resource is especially high (Williamson 1985). Williamson introduced this term as part of the theory of transaction costs.

I have now identified six dimensions for construction projects. Difficulty, complexity, novelty, and resource specificity are independent of each other. However, variability and

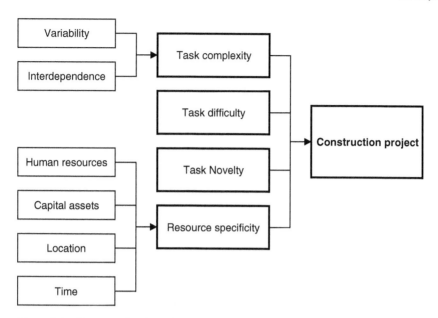

Figure 4.1 Conceptualization of construction projects.

interdependence strongly affect complexity with regard to the number of parts and their interrelationships. Subsuming these two dimensions under complexity, we get a scheme with four dimensions determining a construction project (Figure 4.1).

4.1.3 Typical Cases

Each of the four chosen dimensions represent a continuum with values ranging from "high" to "low." To simplify the typology, I will only consider the extreme values of the continuum. In this way, we can designate the variables by discrete attributes. Difficulty has extremes of very difficult and very easy; complexity ranges from very complex to very simple; novelty spans from very novel to very routine; and resources can be very specific or very generic. An array of four dimensions with two possible values each generates 16 solutions. However, some of these are not possible. It is impossible to combine high difficulty with routine novelty and low complexity. We cannot solve highly novel tasks with ease. Routine tasks do not require specific resources. These practical exclusions reduce the possibilities to seven cases (Table 4.2).

4.1.4 Typology

Table 4.2 allows us to develop the typology by further condensing the information. Generic resources and routine tasks typify cases 1, 2, and 3; they differ only with regard to difficulty and complexity. Taking generic resources as the defining dimension, we can distinguish three types of tasks defining a group of projects. Rule tasks describe the simplest construction projects, where traditional rules are sufficient for a solution (e.g. one-family house). Repetitive tasks differ from rule tasks by their increased complexity; here, managers and

Table 4.2 Existing combinations of dimensions.

	Difficulty	Specificity	Novelty	Complexity
Case 1	Easy	Generic	Routine	Simple
Case 2	Easy	Generic	Routine	Complex
Case 3	Difficult	Generic	Routine	Complex
Case 4	Difficult	Generic	Novel	Simple
Case 5	Difficult	Generic	Novel	Complex
Case 6	Difficult	Specific	Novel	Simple
Case 7	Difficult	Specific	Novel	Complex

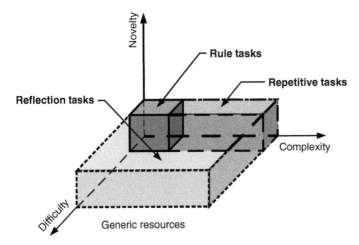

Figure 4.2 R-projects with generic resources.

engineers must apply the rules repetitively (e.g. pipeline). If, along with the complexity, the difficulty is also increased, then we face reflection tasks, where the repetitive use of rules will not generate a solution (e.g. tunnel in a difficult geology). These types form the R-projects (Figure 4.2).

Small companies are well suited to implement rule tasks because such projects do not require a sophisticated overhead. The generic resources are available on the market, the implementation poses no difficult problems, and financing is no hurdle. Such boundary conditions allow for an easy market entry, which is again a reason for the large number of companies in this sector.

For repetitive tasks, there is a need for an iterative approach to the problems and observing impacts of decisions on other parts in order to solve the structural complexity. It is the interdependence of partial solutions that is new, not the general approach. Repetitive tasks provide the natural growth path for companies that master rule tasks.

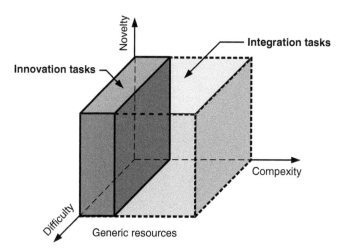

Figure 4.3 I-projects with generic resources.

A further horizontal task enlargement characterizes reflection tasks, and it entails the development of new solution algorithms in planning. This requires reflection based on previous experience. Planning becomes more important, and the necessary engineering resources must be available. A possibility to create a niche market exists based on the competence to solve difficult problems.

Generic resources can also occur together with high novelty and high difficulty, with complexity being the only variable; these are I-projects (Figure 4.3). These tasks are innovation-driven, but they do not require specific resources, with the exception of highly qualified personnel.

Innovation tasks are novel and difficult to solve (e.g. Frank Gehry's Dancing House in Prague). This is the ideal setting for companies with a strong engineering background; they can be of small or large size.

Integration tasks add the dimension of complexity to novelty and difficulty, thus increasing the size of the tasks (e.g. modern sports stadia). In such cases, expertise most likely comes from different sources, and there must be a system leader to integrate all the efforts. The ideal company must have a command of superior project management and engineering competencies. This can be a specialized company or a large one.

Special tasks (S-tasks) combine high degrees of novelty and difficulty with specific resources (Figure 4.4). Specific resources are expensive, and so companies that tackle such problems must possess the necessary capital.

Smaller high-tech buildings pose specific tasks (difficult, novel, and specific), and there are high demands on engineering and financing. An example is the Sydney Opera House.

Megaprojects are examples of select tasks; they have the highest requirements in all four dimensions. An example is the Doha Metro Project with 350 km of metro lines and 100 stations. A company tackling such a task must have many highly qualified managers and engineers with megaproject experience. Access to financing is also of utmost importance.

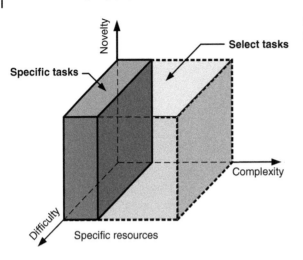

Figure 4.4   S-projects with specific resources.

I have now developed a typology of construction projects by logical deduction. Starting from a review of the relevant business and construction literature, I have identified four independent dimensions. Seven possible combinations exist based on the four dimensions. A discussion of these seven cases resulted in the identification of three project groups:

- R-tasks or routine projects; they are derived from rule, repetitive, and reflection tasks.
- I-projects or innovation projects; they are driven by innovation and integration tasks.
- S-projects or special projects: they are the results of specific and select tasks.

These ideas have been summarized in Figure 4.5.

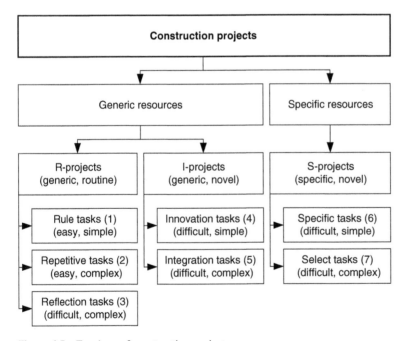

Figure 4.5   Typology of construction projects.

Figure 4.6 Expansion path for contractors.

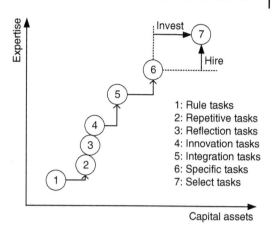

1: Rule tasks
2: Repetitive tasks
3: Reflection tasks
4: Innovation tasks
5: Integration tasks
6: Specific tasks
7: Select tasks

The development for a contractor starts most often at rule tasks and may advance to select tasks. Figure 4.6 shows a development path with an expansion of expertise and capital assets. It is not necessary to stick to this path, but jumping more than one level will prove difficult. Of course, it is possible to buy into technology when enough financial resources are available. It is also possible to attract external financing.

4.2 Complexity of Megaprojects

Complexity is an intriguing characteristic of construction projects. Almost every text on project management or its adjacent fields of studies, such as estimating, scheduling, logistics, or supply chain management, mention the complexity of construction projects (e.g. Bennett 1991, Mubarak 2010, Sullivan et al. 2010). It has become a term like "large" and "beautiful," where evaluation and understanding lie in the eyes of the beholder. Researchers have published scores of articles on complexity; alas, there is no agreement on a definition!

A deterministic definition considers the number of elements in a system and their possible relation. This allows for calculating a complexity value. If n is the number of elements in a system and r the number of all possible relations, then complexity (C_d) becomes:

$$C_d = n + r = n + n * (n - 1)/2$$

If we take C_d to represent a communication system with n actors, then we have r possible communication channels. With 500 employees in an organization, there are 124 750 communication channels. Unfortunately, this does not describe organizational reality. Hierarchy channels communication, tasks, and personal preferences require that some channels are used more than others. In my opinion, this is not a suitable definition for complexity; instead, I will call this complicatedness and assign it the letter K.

Complicatedness (K) of a system = (def.) The number of elements (n) of a system and all possible relations (r) between the elements. We can mathematically determine the value as $K = n + r = n + n * (n - 1)/2$

For a definition of complexity, we need to develop the concept further. Sargut and McGrath (2011) define a simple system by a low degree of interaction and dependable predictability. Complicated systems comprise many elements and many interactions functioning according to clear patterns; they are predictable. On the other hand, complex systems are identified in terms of multiplicity, interdependence, and diversity; in addition, their outcomes are difficult to foresee. The same system configuration at the start allows for different results. Gidado (1996) takes a different approach by concentrating on components (inherent complexity, uncertainty factors, number of technologies, rigidity of sequence, and overlap of stages) and interactions between these. For him, complexity has a purely technical character. These positions represent two ends of a continuum for the definition of complexity: the first is highly abstract and flexible, and the second is concrete and more rigid. Viewing complexity from not only a technical perspective is a rather new approach (Antoniadis et al. 2012). An abstract definition allows for incorporating nontechnical perspectives. The example of a communication system shows that we need to approach the definition of complexity from this end of the continuum.

It seems difficult to define complexity without a framework. A suitable one is the Luhmannian systems theory, which understands the world as unmanageable due to its overwhelming complexity. Therefore, we need to create systems in order to reduce complexity to a manageable degree. Construction projects are one type of a system. Depending on how we draw the system borders of a construction project (one-family home or petrochemical plant), we face a remaining complexity, an eigencomplexity. These are the two technologies of complexity reduction – differentiation and management of eigencomplexity.

Drawing on the definition of Sargut and McGrath and keeping in mind communication systems, I would like to add another dimension to the discussion, i.e. impact. It does matter whether a cause at one point of a system has a large or a small effect on the configuration of the complete system (Wilke 2000); we are facing weighted interactions between the elements (Geraldi 2008).

In the context of megaprojects, complexity is only important with regard to decision-making. A philosophical view of complexity (what is the sense of life?) has little space in the construction office. The whole point of systems theory is creating manageable entities that enable us to make decisions. It is likely that we can improve complexity management if we can identify the nature of complexity and avoid unnecessary complexity (Pennanen & Koskela 2005).

4.2.1 Defining Complexity

Following the discussion above, I would like to put forward a nominal definition of complexity comprising a number of elements, a number of interactions, and the strength of the impact that ripples through these interactions to the elements. This definition is in line with Sargut and McGrath (2011) and a number of other authors. It seems that the efforts to find a definition converge on these three characteristics, with two of them (elements and interactions) being fixed (a given) and the third one under scrutiny.

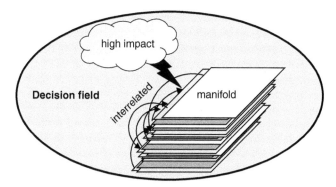

Figure 4.7 Complexity of a system.

 Complexity = (def.) The number of elements, their interactions, and the strength of impacts of a defined system with regard to decision-making. Complexity is a mental construct.

For our purposes, a megaproject is the defined system, and only those elements and interactions that need attention (decision-making) are of interest. In a highly complex system, the elements are manifold and interrelated with some connections of utmost consequence (Figure 4.7).

This definition is a general one that refers to any type of system. For a construct of construction project complexity, we will need to find further determining factors. Complexity does not remain constant over the lifespan of the project; instead, it is dynamic. Therefore, we face different configurations of complexity at different times. The overall aim is to reduce complexity by decision-making, performing, communicating, coordinating, and learning. As complexity is a state of configuration, dynamics cannot be part of the concept of complexity itself. However, a dynamic environment produces different configurations and levels of complexity (Brockmann and Girmscheid 2008). At the beginning, the designer has hundreds of thousands of components available and an infinite number of combinations. For example, there are more than 200 faucets available in most countries, and it is a negligible component for a one-family home. No architect considers these possibilities; they prefer maybe 5–10 and choose without deep reflection. There is no other way to deal with the complexity.

4.2.2 Construct Dimensions of Complexity

As mentioned, the discussion on complexity has progressed from just considering technical complexity to include other categories. Baccarini (1996), for example, distinguishes between organizational and technological complexity. Brockmann and Girmscheid (2008) introduce task, social, and cultural complexity based on Wilke (2000). Task complexity combines parts of organizational and technological complexity, especially planning and organizing. It excludes leadership, which is part of social complexity. Task complexity is highest at the beginning of a project, and it is the job of the project members to reduce it to zero at the end.

There can be little contention that the number and diversity of stakeholders in a project and the strength of their impact (interest and power) increase its complexity (Chinyio and Olomolaiye 2010). I term this "social complexity." Social complexity has a certain level at the beginning and another level at the end. It is never zero, fluctuating according to the relationship of all influential parties in a project.

The same holds true for the influence of culture on construction projects (Tijhuis and Fellows 2012). In all cultural studies, the point is to show how much the stakeholders' cultural diversity influences project outcome. The more the cultures that meet in a project, the more complex it becomes, since it requires the coordination of an increasing number of different cognitive maps; I term this "cultural complexity." Cultural complexity always exists, even if only at the level of different organizational cultures. We need to manage it in order to limit negative and enhance positive impacts.

Two additional forms of complexity develop with time: cognitive and operative complexities. At the start of a project, their complexity level is zero. Cognitive complexity mirrors the differentiation with which we think about a construction project; this increases with time as we understand a project better. Due to the singularity of megaprojects, we do not have a well-developed understanding of project specifics. Project members learn new aspects of their project at all times. They have the best opportunities to comprehend project demands by adjusting their cognitive maps. The more differentiated these cognitive maps are, the better they help project members to deal with task, social, and cultural complexities.

Operative complexity is the degree of freedom available for project members to determine the most appropriate actions. Again, this develops with time through trial and error as well as observation. The more mature the operative complexity, the better are the chances to deal with the complexity. Here, we have the perplexing situation that developing a higher (cognitive and operative) complexity allows for managing or reducing task, social, and cultural complexities. However, we are quite familiar with this phenomenon: babies need to develop high levels of cognitive and operational complexities for independent survival.

A confined space influences task and social complexity. Restricted space for the tasks (e.g. inner-city construction) and social interactions (e.g. interferences with the public) increase these two types of complexity. All five types become more complex when time is short. A dynamic environment also increases complexity, in some cases dramatically. Dynamics require an adjustment of the system, maybe with new borders, new tasks, new risks, and new learning processes (Figure 4.8).

4.2.3 Factors of the Construct Dimensions

So far, I have presented five construct dimensions (task, social, cultural, cognitive, and operational complexities). Different factors can characterize each of the five dimensions. The number of elements, interactions, and impacts are factors of all five dimensions, for they are a constituent part of the general definition.

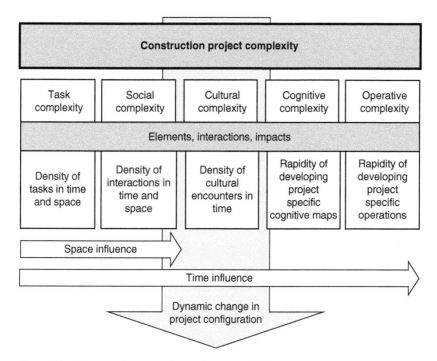

Figure 4.8 Concept of construction project complexity.

Task complexity: The concept of "density" has two categories. The first one applies to decision-making; here, time pressure increases complexity. There is not enough time to gather information and analyze the situation. Space limitations also increase complexity. We only have to think of large numbers of subcontractors working in a confined space. It takes good planning to avoid an avalanche of claims and luck to finish with just a few.

Social complexity: Similar to task complexity, social complexity arises when interactions take place within a short period. Time pressure comes across as chaos because there simply is not enough time for coordination. Space limitations are not a real problem; the opposite is true, i.e. the scattered locations (dispersion) of a project team increase complexity (Dainty et al. 2006). This is definitely a challenge for international projects: large spaces between global project participants breed mistrust.

Cultural complexity: Culture only becomes problematic when different cultures meet. A good guide to measure the differences between cultures is the work of Hofstede and Hofstede (2005). They provide data for many different cultures and use six factors to describe culture: power distance, uncertainty avoidance, masculinity, individualism, long-term orientation, and indulgence. A number of researchers have criticized their approach. However, it remains the best available framework for management.

Cognitive complexity: Everyone in a project comes with different mindsets, i.e. different cognitive maps. Social and cultural complexities capture these differences. Cognitive maps provide us with orientation without the need to analyze the situation from the beginning. The important point is that the people in a project possess the appropriate maps. When a project is singular, we have to adapt our maps to the situation at hand.

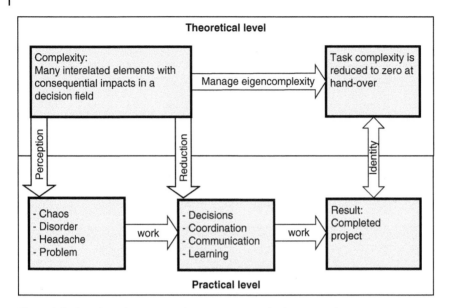

Figure 4.9 Relationship between chaos and complexity in megaprojects.

This is a learning process. However, the maps with which we start the process should be as applicable as possible. Put a little simpler, we need to have the required know-how. Some researchers refer to cognitive maps as frames. Snow and Benford (1988) and Gamson et al. (1992) define three types of subframes: (i) diagnostic frames, (ii) identity frames, and (iii) prognostic frames. These can serve as factors for cognitive complexity. The question for evaluating project complexity helps determine ex-ante whether or not the people who will be decision-makers for the project have access to applicable cognitive maps and whether they are capable of developing these maps in accordance with the project through learning. Decision makers who are not able or willing to learn the required singular approach will increase project complexity. Diagnostic frames help analyze situations appropriately, identity frames help to reduce coordination needs, and prognostic frames allow assessing the future properly.

Operational complexity: While cognitive complexity is concerned with know-how, operational complexity deals with know-that, i.e. the ability to do the correct things. Again, this is a learning process. What is of importance are the available operational skills at the beginning of the project and the ability and willingness to learn. Technical and management skills belong to operational know-that.

Few megaproject participants will state that they are managing the eigencomplexity of the system "megaproject" to reduce task complexity to zero. In their words, they do not perceive complexity but chaos, disorder, headaches, or problems. They decide, coordinate, communicate, and learn, and by doing all this, they slowly finish the project. However, the theoretical and practical levels are interrelated, and the results are identical (Figure 4.9).

4.2.4 Complexity Development

As described, task complexity is high at the beginning of a project. If we assume that the project starts with signing the construction contract, the contractor's bidding team has

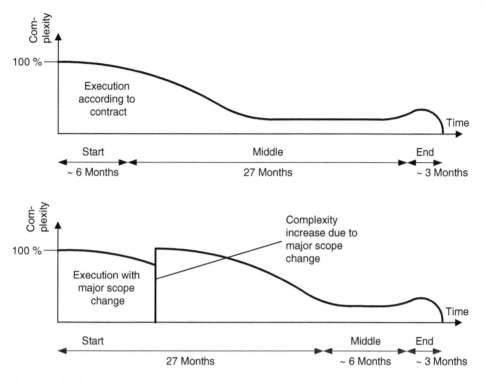

Figure 4.10 Complexity development (project management).

already reduced complexity through the work associated with the estimate. Neglecting this, task complexity is 100% (maximum) when work starts on-site. If a construction period of 36 months, no defects, and no warranty period (just in this model) is considered, task complexity is zero at handover. However, the reduction is not linear. In the beginning, it is very slow. During this time, the project management team makes many important decisions which will show results only later. When these decisions show results, complexity reduction accelerates and approaches zero, but it will always stay somewhat above this value due to effects that the contractor cannot control or owing to mistakes. At the end, the complexity increases again through all the work associated with closing down the project before it abruptly turns to zero. Figure 4.10 shows a curve that is indicative of project management with the above-discussed characteristics. A change order can result in increasing complexity to almost 100% again. The period with high complexity extends almost throughout the project. As productivity is lower at high complexity periods, this has financial consequences. In many cases, change orders impact not only task but also social complexity. Mistakes in planning or execution are the most probable causes for change orders. Finding the responsible party for such mistakes often leads to a deterioration in working relationships (the blame game).

The time given for the start (6 months) and end (3 months) are indicative of a well-run project. Start and finish periods are different for each megaproject.

The run of the curve is different for design and construction. In case of a design/build contract, the curve will approach zero complexity perhaps after a third of the project is through. At that time, all important design problems are clarified. Execution mistakes and design

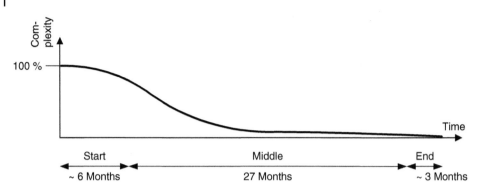

Figure 4.11 Complexity development (design).

omissions keep the complexity lingering above zero. In such cases, as-built drawings must show the actual construction. However, there are numerous construction interruptions that do not affect design. Among them are inclement weather, accidents, traffic interruptions, strikes, or public interference (Figure 4.11).

5

International Construction Management

The Engineering News Record (ENR) reports continuously on construction markets. It publishes a list of the top 250 international contractors ranked according to their turnover outside the country of their headquarters. The turnover of these international contractors in 2017 amounted to 482 billion USD (Tulacz and Reina 2018). This equals roughly the total international market volume. We can also take it as the approximate sum for international works on megaprojects, as local companies will take care of the smaller contracts. We need to add to this amount of 482 billion USD the megaprojects completed in one country by national contractors, such as the Boston Artery.

5.1 International Construction Joint Ventures

The typical configuration for the completion of megaprojects are construction joint ventures, in many cases international construction joint ventures (ICJVs). Joint ventures have the following characteristics:

International joint venture (IJV) = (def.) A minimum of two independent companies join their activities in pursuing a common goal while sharing responsibilities and risks. For international joint ventures, at least one of the companies must have its headquarters outside the country of the joint venture.

Some IJVs are equity joint ventures governed only by a joint venture contract. Others are contract joint ventures (CJVs) with an internal joint venture contract and an additional external contract.

Equity joint venture (IJV) = (def.) A minimum of two independent companies join their activities in pursuing a common goal while sharing responsibilities and risks. An internal contract defines these goals. The partners provide equity to achieve them. Equity joint ventures are typical for exchange goods.

Contract joint venture (CJV) = (def.) A minimum of two independent companies join their activities in pursuing a common goal while sharing responsibilities and risks. An internal contract defines the relationship between the partners and an external contract the goals. Contract joint ventures are typical for contract goods.

Construction companies agree to implement contract goods and the contract with the owner defines the responsibilities; therefore, they form contract joint ventures. Since contractors always produce contract goods, the term "construction joint venture" equals the term "contract joint venture." Thus, we can define international construction joint ventures.

International construction joint venture (ICJV) = (def.) A minimum of two independent companies join their activities in pursuing a common goal while sharing responsibilities and risks. At least one of the companies must have its headquarters outside the country of the joint venture. An internal contract defines the relationship between the partners and the external construction contract with the owner, the goals, rights, and responsibilities. Contractors often form international construction joint ventures to carry out megaprojects.

I stress the difference between equity and contract joint ventures because authors do not always mention it in pertaining literature. This can lead to confusion. While one success factor for equity joint ventures is their longevity, it is brevity for construction joint ventures. If two companies join to develop a cell phone and this phone sells for years and years, this is a success. A megaproject that does not finish on time is a failure.

Figure 5.1 shows four companies from three countries (one is a local company) forming an ICJV. The internal joint venture contract determines the internal rights and responsibilities among the companies (partners). The construction contract between the owner and the ICJV describes the construction task and regulates the rights and responsibilities between the two sides. The partners of the joint venture are jointly and severally liable. In case of problems, the owner can approach the ICJV or any single partner and demand compensation.

Most of the time, although not always, contractors form ICJVs for megaprojects. Two of the most important reasons are the availability of necessary finances and diverse technology. From now on, I will take this to be the case in point: ICJVs are the typical organizational form for implementing megaprojects.

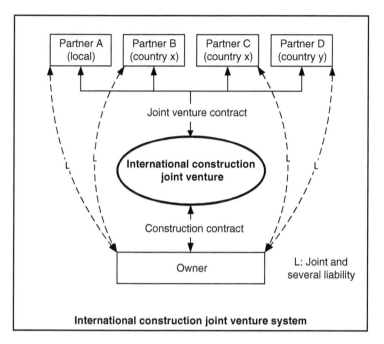

Figure 5.1 International construction joint venture.

5.2 Global Contractors

Possible partners in international megaprojects are global contractors – the big players. Few local companies invite a foreign one to set up an ICJV without good reason. The foreign company must provide qualities to satisfy such reasons. It will have a company history showing growth from a regional to national, then international, multinational, and finally, global level. Work on the global level demands management competence to create successful structures for megaprojects (dealing with high task complexity). In international business, a cultural competence is indispensable because the ICJV will comprise members from many countries (high cultural complexity). This requires turning cultural diversity into an asset instead of a burden. The number of stakeholders in international megaprojects is large (high social complexity). Therefore, well-developed social competence is required. Megaprojects often involve contact with and involvement of highly placed political entities. This might be on special occasions such as the foundation stone ceremony or more regular negotiations; thus, diplomatic tact is essential. In order to be of interest to local contractors, the global contractor needs a matching reputation. Otherwise, it will not draw attention for the first contact. Most of the time, global contractors bring the expertise of cutting-edge technologies. These might include construction or management technologies. Given the size of megaprojects, large financial resources are required.

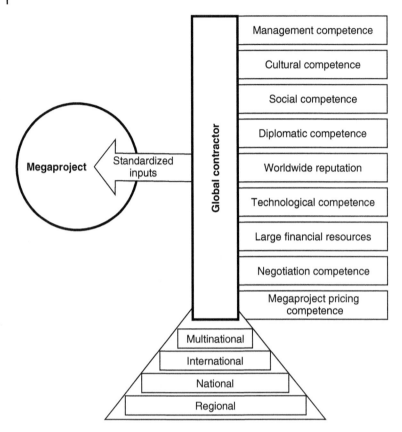

Figure 5.2 Competencies of global contractors.

Assuming a 2 billion USD contract and contractually stipulated bonds totaling 10%, a credit volume of 200 million USD is the precondition for submitting a bid. Negotiations will accompany the tendering and execution stages. These are often highly complex and demand knowledge of law, financing, and technology. Finally, and most importantly, for contractors, global contractors must have the ability to estimate the contract price correctly (Figure 5.2).

Global players in construction do not provide for standardized outputs, as with exchange goods. An iPhone is an example of a standardized output. Global contractors standardize inputs. They can set up a network of contacts in any location across the world and manage it successfully; they are able to create the necessary and successful system.

Typically, a construction company starts on a regional market and might develop into a national, international, multinational, or global contractor. Most companies stay on the regional level; few make it to the top. Small regional contractors can be more profitable than global ones. There exists no causal link between size and relative profitability. This removes a strong incentive for developing into a global contractor. The number of global players is much smaller than that of regional players, and thus competition is smaller, even considering fewer available megaprojects. The ratio of global players per megaproject is typically smaller than for normal projects. However, competition is often strong enough to keep profit margins low.

The listed competencies also apply to the project manager. However, the reputation of the company and its financial strength will provide enough cover for the project manager. Few, if any, megaproject managers have a global reputation by themselves.

5.3 Goals for International Construction Joint Ventures

ICJVs comprise different actors from different companies and different countries. A typical goal difference exists between foreign and local companies. Local companies rely on the local market while international contractors can choose a hit-and-run strategy. For the local companies, the shadow of the future (further contracts) is much more binding.

 Hit-and-run strategy = (obs.): The Taiwan High-Speed Railway Corporation is the owner of the high-speed rail link between Taipei in the north of Taiwan and Kaohsiung in the south. There is no more space in Taiwan for another such project. The owner just had to offer this one project and could not enforce compliant behavior by the ICJVs through further contracts, i.e. through economic incentives. They announced that they would take a strictly legal approach. The ICJVs said they would answer in kind; this amounts to a hit-and-run strategy from both sides. Compromises were not an option. In this case, even the local companies could run away from this special owner.

Goals can be concurrent, competing, or neutral (Figure 5.3). Concurrent goals enforce each other; for example, strict contract fulfillment (goal 1) will lead to keeping the schedule (goal 2). Competing goals contradict each other and require a compromise; high quality (desirable goal 1) entails higher costs (undesirable goal 2). Neutral goals do not influence each other; the concrete quality for the foundations does not affect the organization of the contractor.

Two partners in an ICJV do not have a single but a bundle of goals; these can concur as two-sided goals or compete as one-sided goals. When there are more partners to the ICJV, the system of goals is even more complicated.

Within an ICJV, partners have concurrent and competing goals. Table 5.1 provides some examples (Badger and Mulligan 1995): the foreign partner gains access to a new market (i), he can fulfill the contract requirements (ii), and learn about the market (iii). All these points threaten the local company with new competition. The local partner accesses new technology (iv), he can enter new market segments with this technology (v), and he can serve core customers in these new segments (vi). The foreign contractor enables a possible competitor, maybe to his future disadvantage. From economies of scale (vii), risk dispersion (viii), increased financial strength (ix), and reduced competition (x), all ICJV partners can only benefit.

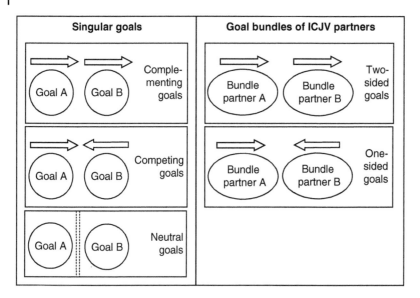

Figure 5.3 Goal configurations.

Table 5.1 Benefits and risks of goals in ICJVs.

	Advantage	Disadvantage
One-sided goals		
Local market access	Foreign partner	Local partner
Meet foreign requirements	Foreign partner	Local partner
Learn local markets	Foreign partner	Local partner
Access to new technology	Local partner	Foreign partner
Enter new market segments	Local partner	Foreign partner
Serve core customers	Local partner	Foreign partner
Two-sided goals		
Economies of scale	All partners	Nobody
Risk dispersion	All partners	Nobody
Increased financial strength	All partners	Nobody
Reduced competition	All partners	Nobody

Goal formulation is a problem not only among partners of an ICJV but also within the hierarchy of the partner companies. The more hierarchical levels there are, the more difficult goal formulation becomes due to competing goals. Figure 5.4 illustrates the possible problems in case there are three levels involved (ICJV, local branch, and board of directors).

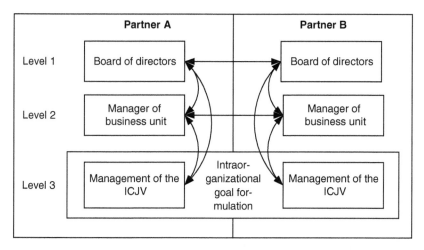

Figure 5.4 Goal formulation in ICJVs.

The partners of the two companies will discuss with each other on the same hierarchical level. There will also be discussions in each company, sometimes omitting one level.

The ICJV management will always have the interest to fulfill the project goals in the best possible way, unless there are incentives from the parent company to do otherwise. Foreign and local partners undergo the process of intra-organizational goal formulation. Within one company, there might be interests that differ from optimizing the results from an ICJV. A branch manager will optimize the results for all projects in his area. In the board of a multinational construction company, the members are typically responsible for different regions. This might generate conflicts among the different hierarchical levels.

We can measure the success of an ICJV theoretically using the following formula (Eisele 1995):

$$I_{TICJVj} = \sum_{i=1}^{n} (S_{ij} * D_{ij})/n_{ij}$$

The denominations have the following meanings:

I_{TICJVj}:	Index of the total success of the ICJV for partner j
n:	Number of goals
i:	Goal i
j:	Partner j
S_{ij}:	Significance of goal i for partner j
D_{ij}:	Degree of fulfillment of goal i for partner j
n_{ij}:	Number of goals i for company j with $S_{ij} \neq 0$

If this formula tells us anything, it is that the process of goal formulation is extremely arduous. I have never seen the use of a formal approach; on the other hand, it is common

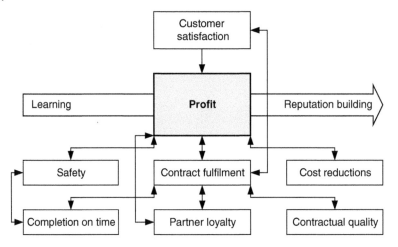

Figure 5.5 Goal system.

to witness infighting to advance the individual goals of one contractor over those of other ICJV partners and between different hierarchical levels of a contractor.

Goal formulation in ICJVs = (obs.): The negotiating team for a megaproject submitted a price that was only 55% of a similar project a few years earlier. The price of the second bidder was almost twice as much. Nobody noticed the differences, and the board members signed the contract. Later, the project manager complained about the price and asked for a reliable estimate as a guideline for the project. All higher levels in the company refused to provide such an estimate because of their responsibility for the signed contract. The project manager had to manage without a reliable budget.

In the end, the board of managers fired all higher-level managers, but by that point, it was too late. Private goals were more important than company goals.

Interviews with participants in ICJVs provide the result shown in Figure 5.5. Goals for ICJVs are interdependent. The paramount goal is profit maximization, and all other goals are subordinated. Cost reductions generally lead to increased profits. Contract fulfillment includes a quality acceptable to the owner and timely completion. If these indirect goals do not serve profit maximization, they will lose importance. This is also true for partner loyalty, safety, customer satisfaction, learning, and reputation building. One interviewee explained that he would spend 2% of the project value on customer satisfaction if and only if the profits were above 6%. This clearly shows the prominence of profits in the goal system.

5.4 Success Factors for Megaprojects

I have already discussed success factor research in Section 3.2. The search for mechanisms in the intersubjective World 3 does not promise better results than a search for Easter eggs

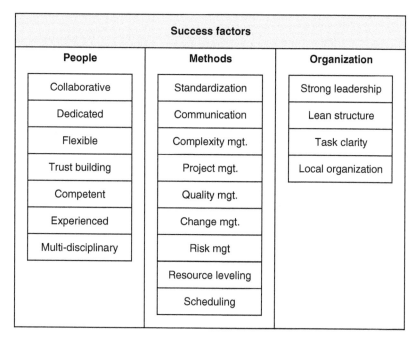

Figure 5.6 Success factors for the design of megaprojects.

at Christmas: they do not exist. What exists are shared cognitive maps of experts. All maps undergo constant updating, and thus, change. The experts develop them in a sensemaking process by connecting them with the physical world of construction (World 1). As such, they are neither chimaeras nor natural laws.

We can expect a certain overlap between a strategic approach and success factors (understood as helpful but not guaranteeing success). Nobody would choose a strategy that does not build on ideas that promise success. However, when I discussed success factors with more than 90 interviewees from global design firms, most of them mentioned factors of the operative level (Brockmann 2013). We can divide these into three categories: people, methods, and organization (Figure 5.6).

On the one hand, the required characteristics of the people working on megaprojects hold no surprises. The following are seen as important personal characteristics: the ability to work in a team (collaborative), to be dedicated (committed), to be flexible, and to be able to build trust. The professional characteristics mentioned are competence, experience, and a multidisciplinary orientation. Nobody mentions a need for specialists. This supports the findings that integration is more important than specialist knowledge. Specialists are more abundant than generalists with sufficient expertise. On the other hand, I am surprised that neither the ability to learn nor a drive for innovation is part of the list. Maybe the interviewees took them for granted?

The most important method is complexity management through standardization and communication. Resource leveling and scheduling are also forms of standardization; neither design firms nor contractors can easily adjust their manpower to the ever-changing requirements of a megaproject. Projects, quality, changes, and risks require management. The only observation here is that the application of risk management is not widely used in megaprojects; in fact, it is very seldom applied by contractors.

Resource leveling = (obs.): The design firm for a design/build road project in Asia started the design with enthusiasm. The ICJV had agreed with the owner about where to start, and the design firm worked on this design package. There was a lot of time pressure, and everybody worked long hours and weekends willingly; at home, the families began to grumble. With pride, the design team presented their first design package. However, the owner had shifted the access to the right of way to another location. The ICJV had to tell the designer to start all over again in this location, and the time pressure did not change. Now, the grumbling arose in the design team as well. Why should they bear the burden for wrong decision-making? After some hesitation, they went to work head over shoulders again, long hours and weekends. The unhappiness in the families increased. Once the second design package was ready, the owner shifted the start point again…

No wonder resource leveling is important.

For the category of organization, the idea of a lean structure with strong leadership for integration is important. Too large an organization makes solving the design and the construction tasks more difficult. In a lean structure, multidisciplinary people are important. Task clarity is more a demand or wish than an observable phenomenon; there are too many change requests during project implementation. One designer mentioned at the opening of a metro that it should be the best 8-year-old metro in the world, when the construction time was 8 years. This is the idea of a design freeze. It would provide task clarity and a minimum of changes. It might also mean lost business opportunities.

 Design changes = (obs.): The design of a metro project in the Middle East was complete when an important person in society wanted to build a new shopping center. He demanded shifting the nearest station to give direct access to his shopping center. Personal interests superseded the good idea of a design freeze, and the project dynamics changed once more for this metro.

I have already mentioned that construction projects rely on local networks. Contractors must always move to the site. Design firms will mostly work in the back office at some other location on the globe. Moving the design team to the site is inconvenient and expensive. Nevertheless, it is important that a strong design manager is locally present.

While the interviewees focused on design, demands for construction are not much different. This holds especially true for personal and professional characteristics and for the organization, less so for the methods. Construction project management requires a more refined approach than design management, since the required resources and the financial investments are much larger.

5.5   Key Personnel

Reviewing the previous results, it becomes obvious that people are key to success. Flyvbjerg (2014) quotes a colleague claiming that if project managers of conventional projects require the equivalent of a driver's license, then megaproject managers must obtain a pilot's license for jumbo jets. Megaproject managers must have the right experience (Figure 5.7). The list of required experiences is long; it seems almost impossible to find a project manager who fits perfectly into the picture. Therefore, we need to prioritize the experiences. Megaproject experience or, in more academic terms, the ability to deal with the overwhelming complexity of megaprojects in all its facets (task, social, cultural, cognitive, and operative) is indispensable. Managers who do not have command of the required skills will make the wrong choices, and the learning process will take too much time.

In second place comes international experience. International projects might not always be megaprojects, but all international projects are large or special. Cultural competence is a key to success in an international environment. In the beginning, local experience may be provided by advisors, but the key personnel must be able to build up local relationships quickly based on their cultural competence.

Management experience comprises project management, design management, and quality management. A special type of experience is system experience – e.g. a railway project with its civil infrastructure, electrical communications and signaling, and mechanical rolling stock. This can be learned on the job, as specialists will have the required interfacing knowledge. Specific design experience then becomes the knowledge base for specialists.

Experienced people are crucial for megaprojects. A balance is required between generalist and specialist experience. The specialists must provide in-depth knowledge in their areas as the design and construction of a megaproject often advance the cutting-edge of our knowledge. However, specialists are not the bottleneck to megaproject success.

The management team must integrate all the different aspects of a megaproject, i.e. task, social, cultural complexity, and cognitive as well as operative complexity. As it is almost

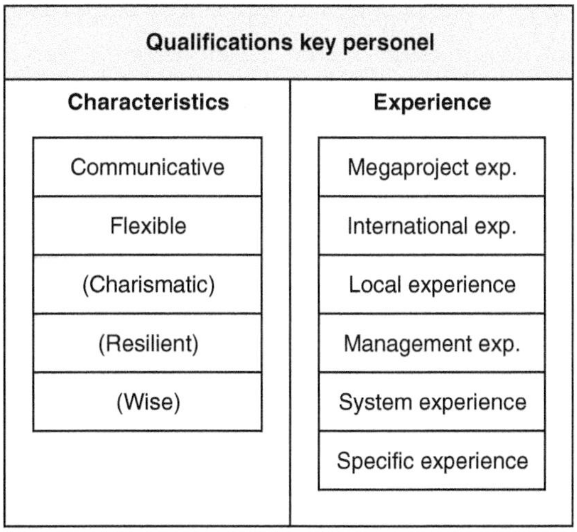

Figure 5.7   Required qualifications for key personnel.

Qualifications key personel	
Characteristics	**Experience**
Communicative	Megaproject exp.
Flexible	International exp.
(Charismatic)	Local experience
(Resilient)	Management exp.
(Wise)	System experience
	Specific experience

impossible to find managers who possess all the required experiences, the management team must couple their available experiences with the ability to learn. The need to learn is not obvious to all experienced managers, but megaprojects demand a respectful approach. Prioritizing the required experiences for the project management team provides the following list:

- Megaproject experience (condition sine qua non)
- International experience (condition sine qua non)
- Management experience (condition sine qua non)
- Local experience (can be learned)
- System experience (can be learned)
- Specific design experience (not required)

The overall conclusion is that it all depends on the right experience. This seems simple, but why do we so often get it wrong? Is it because we do not heed the lessons learned? Or are we not humble enough when taking on the task to manage a megaproject?

Finding a project manager for a megaproject is no easy task. Among the success factors are a local organization with strong leadership and a lean organization. Who should provide this, if not the project manager on site with his team? Accordingly, contractors look for strong personalities with megaproject, international, and management experience. During the interviews, the candidates have to demonstrate their strength and experience. Most often, company managers will choose the candidate who makes the loudest claims with abundant self-confidence.

Once confirmed as project manager, such a person will have a tendency to walk the talk and show everybody his power and knowledge. This is where things go wrong. If a project manager insists on knowing how to run a specific megaproject in detail, he is dead wrong; he has not turned his attention fully toward the focal megaproject. The main characteristic is a manifold singularity, so that each megaproject requires the development of a new approach. To quote the project manager from Section 1.3: "*And so you have a situation where nobody really knows what the other person is doing, why they are doing it, how they are doing it, and even if they should be doing it.*" A megaproject manager must know – by experience – that he must find a new approach for his megaproject. Learning, not knowing, is a prerequisite at the beginning. The project manager quoted in Section 1.3 was a very successful manager on a number of megaprojects across four continents. If he had given the statement during an employment interview, in all likelihood, no company would have hired him.

We can often observe that parent companies change project managers in megaprojects every 6 months at the beginning. Western companies fire them; eastern companies move them to another position. One of the reasons for this is the slow complexity reduction inherent in megaprojects. Another might be the wrong choices made by companies. Many company managers responsible for megaprojects have never managed one by themselves. When they were young, they served as engineers. When they became older, their career might have developed along the hierarchy of the parent company. Thus, they are lacking in first-hand megaproject experience.

5.6 Expatriate Life

The life of expats is thrilling and demanding. Thrilling because the construction sites are often in locations where tourists like to travel. Unlike a tourist, however, the expat gets to partake in the life of a different culture. When problems arise, he cannot simply pay and leave. The expat has to follow through on the problems and find a solution. If traveling is a learning experience, then expat life is a much deeper one.

Agrawal (2018) says of her part in the design of "The Shard" in London that it was a once-in-a-lifetime chance. Incidentally, we can say this about most megaprojects. Some expats even move from "once in a lifetime" to twice and thrice in a lifetime. What can be better for a full-blooded civil engineer than participation in megaprojects? This is thrilling and demanding: A 60-hour workweek is usually part of the contractual agreement; this does not include overtime. Work on Saturdays is very normal. During the peak construction period of a megaproject, I used to work 90 hours per week. Korean engineers tend to add even more hours; 7 days a week from 08:00 a.m. to 10:00 p.m. are typical. The reward is personal accomplishment paired with high salaries. The terms include comfortable housing close to the construction site.

There are two basic conditions for expat life: it can take place in the middle of a city or in a camp far away from urban centers. My personal experience includes years in New York, Copenhagen, Bangkok, Doha, and Cairo. Who would not want to spend time in any of those cities? Other examples also include a camp on the Sinai Peninsula, with ISIS terrorists not far away, as well as camps in the subtropical Northern Territories in Australia, beaches in Indonesia, the Mekong Delta in Vietnam, the rice paddies in Taiwan, and the desert in Algeria. For many, life in a city is easier.

There is an old saying that 2 years overseas allow for half a house and/or a broken family. Families and relationships can suffer or profit from expat life. Companies sometimes encourage bringing families along, because it allows for mental stability. On the other hand, it costs considerably more money. The costs for private international schools are especially high. Host countries often do not allow spouses to work. They must be able and willing to organize a satisfactory social life, including voluntary work. Children have to move to a new school and make new friends. The opportunities and the exciting life are the reward. We decided to move about as a family, and the result is a rich and close relationship. Often, expats also meet spouses in far-away countries. It is always amazing to see the diversity of marriages on megaprojects. Many of these marriages last; others break up.

Finally, there are abundant stories about adventures, sex, and crime. It is easy to fill pages with such stories. I know, for example, of threats by drug cartels in Colombia (a pistol to the head for passing the car of a drug lord), of getting lost – and found again – on trips into the desert (Oman), of getting killed for a love affair (Thailand), of imprisonment and confiscation of passports (Qatar), and of abductions (Nigeria). Expats recount these stories over beer or wine in the evening; they are part of life and legend, but occur much less frequently than the repetitions make us believe.

In sum, expat life is rewarding for those with an open mind who dislike daily shifts in the same office for a whole working life.

6

Megaproject Phases and Activity Groups

Analyzing a megaproject with regard to time allows us to create phases or activity groups. We should always remember that any such structure is a product of our minds and never a physical object. The interfaces are not clear cut. If we conceptualize more or less a sequence of events, we can create sequential phases. If we cannot arrange the events in sequence, then we face overlapping process or activity groups.

When an owner decides to pursue a megaproject, he has many contractual options. The first question is how he wants to organize the market. To this end, he can choose between open or selective bidding and single sourcing. Open bidding provides for the maximum amount of competition, as all interested contractors can submit a bid. When opting for selective bidding, the owner can control the quality of the interested contractors based on their qualifications. Owners of megaprojects often organize prequalification sessions for this purpose. Only qualified bidders receive the tender documents; competition is more limited in selective bidding than in open bidding. In the case of single sourcing, the owner decides to discuss with only one contractor. The most frequent approach in the case of megaprojects is selective bidding.

The second question is, how the owner wants to deal with interfaces. The two most important interfaces are the ones between design and construction and between construction and operation. There is a multitude of possible approaches (Masterman 2002). Three procurement contracts are typical (Figure 6.1). Design/bid/build contracts feature two distinct interphases, one between the designer and the contractor and another between the contractor and the owner. Design/build contracts eliminate the interface between designer and contractor. The contractor becomes also responsible for the final design. Design/build/operate contracts exhibit no interface for a certain time. The contractor is responsible for the design, construction, and operation (for a limited time, often for 30 years).

A prequalification for selective bidding and a design/build contract serves as a case in point for the development of the following megaproject phases; the corresponding interphases are rather easy to distinguish. We can understand them as gates where somebody controls the results of a previous phase and allows passage only when the results meet expectations.

1. Project idea: The owner has to have a project idea and find financing for the idea and a construction site. There are many ideas but not enough financing around the world.

Advanced Construction Project Management: The Complexity of Megaprojects,
First Edition. Christian Brockmann.
© 2021 John Wiley & Sons Ltd. Published 2021 by John Wiley & Sons Ltd.

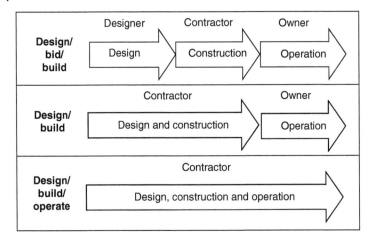

Figure 6.1 Basic contractual options for the owner.

Thus, financing is often crucial. The United Nations catalogues countries into three classes: Developed economies, economies in transition, and developing economies. Especially, economies in transition and developing economies have a strong demand for large-scale infrastructure projects but do not always have the necessary funds.

2. Project development: The owner will by himself or with the help of engineers and economists develop the project to the next maturity level. An important part of this phase is a feasibility study, besides a business plan.

3. Project design: The owner has now the choice to opt for a design/bid/build or a fast-track approach. In the first case, he will advance the design to its final stage before contacting the market. In the second case, the owner will provide only a partial design to the contractor when signing the contract. If he chooses to keep full control of the design, we talk about concurrent engineering, i.e. design and construction overlap. He can also ask the contractor to finish the design, which signals a design/build approach. In this case, the owner must provide at least a conceptual design.

4. Market contacts: When the owner publishes his intention to implement a project, he contacts the market. Next, interested contractors apply for consideration in selective bidding. The contractors decide at this time whether to commit themselves to the project alone or whether it is advisable to form an international construction joint venture (ICJV). In the case of megaprojects, the contractors must demonstrate special skills and financial strength. Owners organize a prequalification for this purpose; they must also complete the tender documents during this time.

5. Bidding period: The contractors or ICJVs work out solutions for the project by working through the tender documents and by submitting a formal bid. The owner often demands a two-envelope bid: One envelope contains technical details of the proposed solution and the other contains the financial terms of the bid. The contractors hand in their bids at the submission date.

6. Contract negotiations: The owner determines the preferred bidder and starts negotiations. Sometimes, an owner negotiates with several bidders at the same time. Both parties conclude the negotiations with the contract signature.

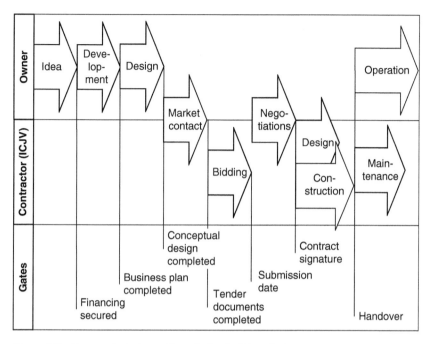

Figure 6.2 Megaproject phases for a design/build project.

7. Construction: The contractor or the ICJV of a design/build project finalizes the design and concurrently starts implementing the project. He must complete construction before handing over the project to the owner.
8. Maintenance period: The maintenance period starts with handing over of the project to the owner. The contractors are still responsible for any defects during this period. At the end of the maintenance period, the owner assumes full responsibility for the project.
9. Operation: At handover, the responsibility for the structure passes over to the owner; he will make full use of it by starting operation. Operation and maintenance periods overlap until the maintenance period ends.

Figure 6.2 depicts these phases in sequence but without indication of the duration of a phase as these vary greatly from project to project. For example, project development takes sometimes 20 years and in other instances 20 months.

6.1 Project Idea and Project Development

These two phases form the so-called front end of megaprojects. It is very difficult to amend any mistakes made at the beginning. This has fascinated many authors (Miller and Lessard 2000; Flyvbjerg et al. 2003; Scott et al. 2011). During these phases, the owner defines the institutional frame for the megaproject. Political, legal, economic, ecological, and social considerations require many decisions. The feasibility study and the business plan also require civil engineering input, but this is not a determining factor.

This book is about construction project management, and the front end is not the primary focus. However, contractors must be able to analyze the institutional frame for its impact on design and construction.

 Front end = (obs.): The two German states of Berlin and Brandenburg planned the construction of a new airport southeast of Berlin. They organized the project as a public–private partnership (PPP), where consortia of contractors could submit bids for the design, construction, and operation of the airport. Two consortia analyzed the project and estimated the cost at approximately 4.5 billion euros. The owner did not like the offers, considering them to be too expensive. They changed the institutional framework and took direct control. The prime ministers of Berlin and Brandenburg became the owners. They insisted on tendering different works in small packages to many contractors. Construction started in 2006 and was supposed to end in 2011. The estimate yielded a total cost until operation of 1.9 billion euros. In 2019, the owners predicted the opening of the airport for fall 2020; the latest cost estimate at completion is more than 7.3 billion euros.

Some of the problems were certainly due to the cumbersome institutional framework.

The lessons learned from the airport Berlin-Brandenburg are not as simple as it seems. Certainly, the institutional framework was not optimal. However, it remains uncertain whether a PPP project would have been more successful. Many design and construction mistakes are also responsible for the delayed opening. These are hardly problems on the strategic level; they are operational problems. Operational problems are part of construction project management from the owners and the contractors' side. The example shows how strategic and operational problems interact with each other. Deeming one more important than the other is the gravest institutional mistake to commit.

What phase is more important is most likely one of personal preference. A football team can serve as an analogy. The team owner decides the strategy and provides financing; he develops a vision and a mission for the team. The head coach with his staff thinks about the tactics and provides blueprints like a designer. He also guides and supervises the team; however, sometimes, he finds his influence limited. The players are on the pitch, and they have to implement strategy and tactical approach like the contractors; they score the goals and have plenty of freedom regarding how to do it. My personal preference is for the pitch and scoring goals; I respect coaches and owners.

6.2 Design Phases

It is helpful to understand the design contract (and later the construction contract) as a principal/agent relationship. The owner is the principal and the designer the agent in a design/bid/build contract. In the case of design/build, the contractor takes over the role as principal. The principal/agent theory advises controlling the action of the agent. The owner

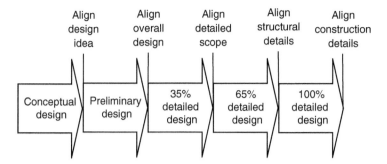

Figure 6.3 Design phases and alignment gates for design/bid/build contracts.

(or contractor) can do this at the end or in between at any point he considers convenient. It is definitely helpful for both parties to introduce controls in between to avoid thrashing a complete design at the end; intermittent checks can minimize waste. We create design phases by these principles. The principal controls at the intermediate checkpoints focusing on progress, costs, quality, and design; a contractor will also pay special attention to constructability. It is clear that they are not mere quality gates but in a more encompassing way alignment gates. The principal controls and corrects results to ensure that the design allows him to reach his goals.

Practice has shown that five phases and four alignment gates suffice in many cases. However, any principal can demand what suits his own special purpose. In between alignment gates, most owners/contractors will organize regular design meetings for fine-tuning. Design is a sequential process in which we add more details at every step. The sequence leads to the label "phase." Figure 6.3 illustrates design phases and alignment gates; I will discuss this point in Section 8.1.

Passing alignment gate 1 (align design idea) ensures that both parties agree on the design idea. Typically, the owner has an idea at the beginning from a business plan or a contract that the designer develops further providing expertise and additional innovation. The design might still be sketchy, leaving space for misunderstandings. Passing through alignment gate 2 (align overall design) guarantees agreement on the general scope. At alignment gate 3 (align detailed scope), all scope details find acceptance by the owner, and the shape of the structure is clear. Alignment gate 4 (align structural details) serves to provide consensus on achieving stability and usability; it does not deal any longer with function and form. Alignment gate 5 (align construction details) serves to reach accord on constructability with all the information that the contractor needs for implementation.

In design/bid contracts, the owner will provide a conceptual design and the contractor will start with the preliminary design.

6.3 Market Contacts, Bidding Period, and Contract Negotiations

6.3.1 Market Contacts

Owners must have a good idea of the players and the situation in the construction market. Megaprojects are subject to close public scrutiny, and often, the press discusses the project

Table 6.1 The 10 top international contractors, 2018.

Rank	Company	Country
1	ACS, Actividades de Construccion y Servicios	Spain
2	Hochtief	Germany
3	China Communications Construction Group	China
4	Vinci	France
5	Strabag	Austria
6	Technip	UK
7	Bouygues	France
8	China State Construction Engineering Corporation	China
9	Skanska	Sweden
10	Power Construction Corporation	China

development. Public work projects require public approval in many countries, and it is very risky to hand out tender documents before approval. In sum, megaprojects announce themselves to interested design and construction firms through their publicity. These parties use their awareness to contact the owner, and it is advisable for him to engage in discussions. This allows the designer and contractors to prepare for the project and the owner to scan the market. Construction supply depends as all other goods on the business cycle. The main characteristic of a construction boom is the full use of resources. This is a difficult situation for owners. Recessions are hard times for designer and contractors; increased competition drives down the prices.

Selective bidding is typical for megaprojects, and owners must find suitable contractors. The size of contractors is one indicator for interesting companies. Engineering News Record (ENR) provides every year a top list of the 225 top international contractors (Tulacz and Reina 2018), which ranks companies according to their turnover in foreign markets. This is a good databank for market research. Table 6.1 lists the top 10 international contractors in 2018.

The top lists by ENR do not only provide information of overall turnover. They also give details about specializations. Further details are available through business reports on the Internet. Once the owner has a list of potential contractors, he can call for a pre-qualification. To this purpose, he invites selected companies and must prepare criteria for the evaluation. Financial data, past experience, and reputation are important. The owner must be aware that reference projects demonstrate past ability of a contractor; they do not guarantee present ability and motivation. He must also take into account that each contractor will show a presentation with plenty of sunshine and blue skies. The owner must analyze in what way he can avoid being dazzled by all the sunshine to get relevant information.

Prequalification = (obs.) An owner for a megaproject in the Middle East took a threefold approach for the prequalification. First, he demanded upfront financial data, which he could countercheck with publicized business reports. Second, he allowed the contractors to make a presentation. Third, he prepared a questionnaire demanding answers to typical megaproject problems. He handed this questionnaire to the team present for prequalification and gave them 30 minutes to prepare answers. Not surprisingly, the questionnaire provided the best insights. Some contractors chose to send a marketing team, which had problems finding suitable answers. Others sent a mix of experienced engineers and managers, and they had better answers. The owner understood the team composition by the contractors as indicative for their motivation. A prequalification is more than a sales pitch.

The contractors must use the time to find out whether they want to tackle the megaproject alone or jointly in an ICJV. The decisive factor is often the monetary project volume. Owners typically require bonds. A bid bond assures that the contractor will sign a binding contract if the owner decides to award him. A performance bond becomes due after signature and guarantees performance according to the contract. Both bonds combined can add up to 20% of the contract sum, if the owner so decides. For a 2 billion USD project, this amounts to a credit of 400 million USD for the contractor. Few contractors can provide such a credit line. This means that the owner predetermines the contractors' organization when setting the size of a project lot. A project lot of 2 billion USD often creates ICJVs with one or two international partners and two or three local partners. Very often, owners forbid contractors to form joint ventures after receiving the tender documents or after prequalification. If they would allow the formation of new ICJVs at this stage, they might face a very limited competition, e.g. six groups could decide to form just two ICJVs. This cannot be in the interest of the owner.

6.3.2 Bidding Period

During the bidding period, contractors often assemble teams of around 20 specialists to work through the tender documents. Most important are legal expertise, innovation know-how, megaproject estimating, and scheduling experience as well as technical understanding. For design/build projects, they have to advance the design to such a stage that confident quantity take-offs are possible. Most companies will cooperate for this purpose with an international design firm. The bidding team has to square 11 partial plans (make-or-buy decisions, scheduling, estimating, logistics, legal organization, health, safety and environmental organization, management organization, site installation, construction technology, resource planning, and quality planning). I will discuss this later in more detail in Section 8.3 (production planning).

In most cases, an ICJV undertakes the tasks described, and the members seldom have the same opinion. Internal discussions must end in a good compromise.

 Bidding = *(obs.)* An ICJV in Asia prepared a bid for a metro project. The owner demanded that the contractors should accept the risk of the soil conditions. This is unusual as the owner remains in possession of the construction site. Soil exploration borings were scarce, and the risk was sizable. The members of the ICJV discussed the topic in depth in their last meeting. They finally decided to accept the risk because, otherwise, they would have lost their chance to win the bid. They also considered the risk in their estimate. The top manager of the ICJV's sponsor had to write the accompanying letter and prepare the bid documents for submission. The other members of the ICJV were shocked to find after submission that the letter stated a refusal to accept the soil risks. The owner excluded the ICJV from further proceedings. Consequently, there was one serious competitor less for the project. What might have been the motivation for the responsible manager? He claimed to have made a mistake stating "we do not accept" instead of "we accept." Maybe he received a bribe from a competitor? This is in any case an interesting application for the principal/agent theory with the top manager as an agent.

The bidding process does not come cheap. Anecdotal evidence shows that four ICJVs bidding in 2015 for the Femern Belt Tunnel between Denmark and Germany spent around 20 million euros each to prepare the offers.

6.3.3 Contract Negotiations

Contract negotiations seldom finish quickly. The submission date for the Femern Belt Tunnel was September 15, 2015, and on May 30, 2016, the parties signed two contracts. It took $8\frac{1}{2}$ months to conclude the negotiations. This was a well-prepared process; in other instances, the period has been longer. The negotiations also cost money. On average, 10 managers and engineers shuttled every week back and forth between Germany and Thailand during negotiations for a megaproject. Representatives of the local partner joined them. Work continued in the back office. To this, we must add the expenses of the owner.

The parties discuss many topics during the months of contract negotiations. Typically, one party writes minutes of meetings and the other party signs them off. Not everything discussed finds its way into these minutes. During this time, a contract understanding develops. This understanding rests on two foundations. The owner has a tendency to believe all problems are solved to his benefit. The contractors like to enforce this belief; they want to conclude the negotiations. On the other hand, they do not want to commit themselves to any additional costs. Both parties in consequence do not share the understanding; there often remain a number of conflicts, which spill over into the construction period.

The (rather precisely) written contract, the (not so precisely formulated) minutes of meetings, and the (quite obscure) mutual understanding set the starting point for construction. When the bidding team hands their knowledge to the project team, it is important to transfer all knowledge. Unfortunately, this never works perfectly.

The discussion of activities up to this point in time do not belong to construction project management in a restricted sense, which only starts after signing the contract. However, it provides vital information.

6.4 Construction and Maintenance

Megaprojects are singular and consequentially innovative. Contractors must apply new approaches; many of the activities are not tried and tested. The Project Management Body of Knowledge (PMBOK 2017) does not define phases but process groups. These are not sequential but overlapping or concurrent. I will describe these in more detail in Section 8.3 (project management); the process groups are initializing, planning, executing, monitoring, and controlling as well as finishing. The five process groups do not suffice for the conceptualization of activities in megaprojects. In addition, they do not describe construction, and specifically, lack detail and context.

Therefore, we need to develop specific process groups for megaprojects. I will call them activity groups to differentiate them from the process groups in the PMBOK. After signature, the contractor must immediately begin mobilization. During this time, he concentrates on production planning, which will be the focus of Section 8.3. Based on the status from the bidding period, engineers and managers on site must develop the solutions further and create a consistent and detailed planning system. Important and well-known parts of this system are budgets, schedules, resources, and construction technology. The second activity group deals with procurement of managers, engineers, and labor as well as material, equipment, and subcontracts. Some equipment and materials are locally available, others only globally with a long lead time. Some equipment is made-to-order and needs design, fabrication, and shipping. A lead time of half a year for specific assets is rather the norm. The third activity group includes testing the construction technology with all the specific assets. Here, the duration of activities will become shorter because of the learning curve. A few repetitions (maybe up to 10) typically reduce activity time by one-third. Depending on the complexity and novelty of the construction technology, it might take (much) longer. Activity group four centers on learning to master the chosen technology. The following activity group five focuses on stabilization of the construction activities. Routine characterizes activity group six. The final one, activity group seven is demobilization.

In sum, we can conceptualize the construction of a complex megaproject as a model with seven activity groups:

- Activity group 1: Planning to align the partial plans
- Activity group 2: Procurement of human resources, materials, equipment, and subcontractors
- Activity group 3: Testing the construction technology (learning curve)
- Activity group 4: Mastering the construction technology
- Activity group 5: Stabilization of all construction processes
- Activity group 6: Routine processes
- Activity group 7: Demobilization of the project

Asset specificity = (obs.): Most tunnel projects using tunnel boring machines (TBMs) as construction technology require specific TBMs because soil conditions vary from place to place. The contractor needs to define his requirements, and the producer will design the TBM with his cooperation. The producer must then procure the materials (steel, equipment, and IT) and build the machine. He ships the TBM in parts to some port on the globe, and the contractor receives it and stores it on site. There, the contractor needs to prepare a starting pit and assemble the TBM in the pit before testing the machine; the learning curve starts with boring the first meters of the tunnel. After mastering and stabilizing the tunneling processes, ideally, a long routine period sets in, followed by a short demobilization.

Figure 6.4 illustrates the overlapping of the activity groups. The first half of the graph shows a wild mix of five activity groups with a rather short duration that require shifting attention. This is the messy part of the project with an overwhelming workload when nobody knows where to start. Constant attention to repetitive details determines the second half (routine) with a bit of additional excitement at the end.

The description of the activity groups is rather technical. Social dynamics accompany and influence the technical contents of each group. Team formation starts immediately after signature. Many owners request the curricula vitae of the key personnel of the contractor as part of the tender documents. The contractors must identify the key personnel prior to signature so that they are available on day one. People from many different countries come together especially in ICJVs, thus creating cultural complexity. All the people joining the ICJV face and develop cognitive and operative complexity. For these reasons, I will split the following subchapters into the categories of task, social, cultural, cognitive, and operative complexity. A design/build project serves again as case in point.

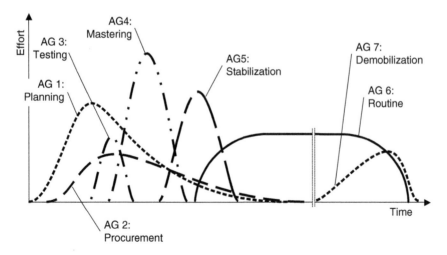

Figure 6.4 Construction activity groups.

6.4.1 Planning and Procurement

Planning and procurement develop concurrently with a little head start for planning. For this reason, I will treat them together. A rather small group of managers and engineers develops the planning and procurement processes at the beginning of a project. This initial engineering group will contain future key personnel and some specialists. It is wise not to add to the overwhelming task complexity at the beginning a large social complexity. The initial engineering group might start with 10–20 persons.

The initial engineering group must design the internal management structures and the product and processes. The product design starts with the conceptual design from the owner and further developments are made during the bidding and negotiation periods by the bidding team. The design determines the construction technology (and vice versa) and the required resources. The planning must reach a stage that the ICJV can draft, negotiate, and sign procurement contracts with confidence.

6.4.1.1 Task Complexity

This is the beginning of the implementation of a project and marks the point of no return – with the exception of demolition. There is a saying that planning of a structural member stops only once the concrete is cast. It is also the time to reconsider and to advance all previous decisions. The number of elements to choose is very large, and the combination of these elements approaches infinity. Starting with the rules of thumb, managers and engineers develop the final solutions. Satisficing behavior is evident, and experience and intuition play a large role. The available information is far from complete. Experience can lead to conservative (this is how we always do it!) or to very innovative solutions, if a large experience serves as a base to charter a path into the unknown.

A schedule can serve as an example. The schedule for the bid might contain 1500–2000 activities. Engineers take it together with plans and the contract to develop a work breakdown structure with tens of thousands of work packages. They rearrange them next into almost as many activities. The schedule must reflect all constraints from the other partial plans. It is clear that the development of the schedule is not a straight process but iterative and interrupted, as information from some other plan is missing.

Procurement is also a hefty task. If we consider a megaproject in a developing economy with low wages, than procurement of materials makes up 70–80% of the project volume. Assuming the project volume to be 2 billion USD, this will add up to 1.5 billion for material procurement. Thus, procurement and management of materials become priority for planning. In developed economies, this is different. Labor costs make up the largest percentage of the project cost (approximately 60% for building), and the focus centers on minimizing labor input.

In sum, the task complexity is the highest at the beginning of implementation when planning and procurement overlap. The owner deals with more and wider-reaching decisions at the front end, but the degree of detailing is limited. The contractor's bidding team has to take into account the decisions by the owner and has to fill in the remaining void, which is large. Again, the detailing is limited. The project team has to consider prior decisions such as those fixed in the contract. It still has a considerable amount of choice and the detailing becomes minute. Task complexity is the product of the number of decisions times the degree of detail for each. It is largest during project planning after signature, especially for design/build contracts.

6.4.1.2 Social Complexity

With its inception, the typical team formation processes start, i.e. forming, storming, norming, and performing (Tuckman 1965). Of course, this is a simplified model but has much explanatory power (Figure 6.5).

The partner companies to an ICJV second the personnel to the initial engineering group. They find themselves thrown together in an office all of a sudden. Almost everybody is excited about the challenges of the megaproject (once in a lifetime!) and keen to meet others. During the phase of forming, managers and engineers will get to know each other. Most will try to assess the others carefully. However, first conflicts arise immediately. The partner companies agreed on project manager, construction manager, design manager, and commercial manager. Some in this group might not accept the hierarchy.

Assessing the project manager = (obs.) The members of the initial engineering group came together in a project. The CVs of the key players were part of the tender documents and, therefore, accessible to everyone in the group. The construction manager and the design manager had a long history with the company, top-level experience in megaprojects, and many years of experience in the country of the ICJV. The project manager did not have comparable experience. The construction manager and the design manager started fighting with the project manager immediately. The decision simply did not make sense to them. The project manager also wondered about it.

Later, the three established a working relationship without becoming friends. Still later, the sending company replaced the construction and design managers.

The transition from the forming to the storming phase is fluid, but there will in most cases come a time when storming intensifies. It is a fight with no holds barred, and it

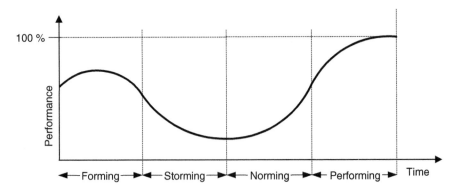

Figure 6.5 Team formation process.

leaves wounds that take time to heal. There are cultural differences related to how people carry out the fighting. Westerners might hit directly, and Easterners indirectly. Arabs might accompany the fighting with a lot of talk; Asians might prefer dealing quietly: The fighting, however, is universal.

 Naming the project manager = (obs.) The partner companies in a megaproject could not agree on a project manager. They send two potential candidates to the initial engineering group. Everyone in the group tried to understand the hierarchy and the situation. The two managers fought each other without mercy, spending more time on consolidating their position than on getting work done. Groups formed around each one to carry out work. When finally the partners named one project manager, suspicion and bad feelings continued. When the partners fired this first project manager 6 months later, the second candidate replaced him. He still carried the burn marks of the negative campaign. The partner companies replaced him half a year later. This is not a good example.

Most of the time, the fighting stops because the demands of the megaproject require full attention and smooth cooperation. The team has to choose between working together or being hanged together. As hanging is no fun, most people opt for setting aside their differences and concentrating on getting work done. Fortunately, there is so much work that it is possible to carve out large areas of responsibility for everyone in the top management team. Only those who place more importance on their place in the hierarchy than on responsibility and achievement will keep on fighting. They need to be sent home.

A normal ICJV will move toward the stage of performing before the end of the planning and procurement phase, which often lasts around 6 months.

6.4.1.3 Cultural Complexity

Many ICJVs incorporate 20–30 different nationalities. Important are the cultural differences between the partners, as they determine the course of action. I will discuss this in detail in Chapter 12 (titled "Cultural Management").

Members of an ICJV typically enjoy cultural diversity. It is one of the reasons why they seek such work. This refers to the cultural complexity on the individual level. Very colorful marriages, partnerships, and friendships are the result. While it is great fun to hear stories of life and work from around the world, it can also be the source of misunderstandings. As I have said in Section 5.6, life in ICJVs is thrilling and demanding; it provides joy and demands extra work.

Cultural diversity on the organizational level requires more attention and planning. However, experienced international managers are able to derive many benefits by creating matches between national characteristics and tasks.

Most ICJVs are able to reduce cultural complexity during the planning and procurement phase considerably. What remains are disturbances on an individual level or different interpretations of situations that arise.

6.4.1.4   Cognitive Complexity

Cognitive complexity describes the specific understanding that the top management team develops for the project. Most of the time, the team starts at zero. During the kick-off meeting when information passes from the bidding team to the project team, the understanding of the project team jump-starts from the outside. From this point onward, the team will dedicate all its energies to the project and slowly start to develop a specific understanding. The singularity of each megaproject requires a very specific understanding. The wider project team, including the owner, is the only group who can develop it. All others just spend too little time and accordingly too little thought on the megaproject. After a while, the top management team will have much, much deeper knowledge and understanding than the joint venture board or anyone else in hierarchy of the sponsors.

During the planning and procurement phase, cognitive complexity increases slowly. It will advance until the end of the project.

6.4.1.5   Operative Complexity

Operative complexity describes the freedom of action by the top management team. In the same way as cognitive complexity, it is zero at the beginning. During the kick-off meeting, the bidding team explains what the next actions are that they have agreed to during the negotiations. From here on, the project management team must chart the way into the future.

As the project team develops understanding, it will also develop simultaneously a course of action that slowly expands the path laid out by the bidding team. During the planning and procurement period, the project team is under scrutiny; it must earn the trust of all supervising managers and assure them that the decisions made and the actions taken will bring success. Once the project team has earned their trust, the freedom of action increases strongly. However, it is limited by the understanding of the problems.

 Partner demands = (obs.) In an ICJV with two partners, the companies' managers asked the project manager to implement their respective accounting system. The two systems differed. This is a typical and unsolvable problem at the beginning of an ICJV. There are many such demands. The project manager has to somehow comply with such wishes (bosses are not always right, but they are always bosses!) and still find the best solution to his own problems.

6.4.2   Testing the Construction Technology

If we take a tunnel-boring machine (TBM) as an example, the project team must first develop specifications, then write a contract with a producer, and sign it. The producer will build the TBM and ship it to a port. The ICJV organizes the transport from the port to the site, the storage, and the assembly. After testing the mechanic, electric, and hydraulic systems of the TBM in the start pit, tunneling will commence. The first tunneling meters belong to the testing phase because we can learn about a TBM only from the interaction between tunneling crew, machine, and soil.

Figure 6.6 Tunnel-boring machine. Source: © Herrenknecht AG.

The planning and procurement stage covers the actions until the delivery of the TBM. The following activities belong to the testing period. It ends when the TBM has reached an acceptable progress for the first time. Depending on the soil conditions, this is often a daily progress of 10–12 m.

With regard to time, a TBM is a special case. Writing the specifications, developing and negotiating a contract, and signing it might take half a year. Producing and shipping the TBM take approximately half a year. Transporting the TBM to the site and assembly in the start pit require 3 months. On sum, the actual testing can begin only after some 15 months. Compared with the starting pit, which must be ready for TBM assembly, tunneling itself starts late. The construction of the starting pit might begin 6 months after signing the contract.

Figure 6.6 shows a tunnel-boring machine, which might cost 15 million USD depending on machine type, diameter, and specifications. It is highly complex tool. The picture conveys the complexity of designing, producing, and operating such a tool, especially together with varying soil conditions.

During this period, the learning curve plays an important role. Due to the learning curve, we can handle repetitive tasks faster. The reason is that, in the beginning, the tools we use might not be complete or in a perfect shape. We must learn how to use the tools efficiently. Empirical data show that the learning curve for the use of column formwork ends after six or seven cycles. It takes more cycles for a complex machine like the TBM to achieve average speed. Complexity determines the duration of the learning curve, and the reason for increased productivity is the better handling of equipment or tools. Other and additional causes drive the progress curve, which I will discuss in Section 13.3.

6.4.2.1 Task Complexity

Specifying, designing, ordering, and producing a TBM reduce task complexity greatly during the planning and procurement activity groups. There might be some late adjustments on site, but they will not be considerable. This can change when the owner orders the TBM, and the contractor wishes to make changes based on his experience. All this is never of a very high complexity. The assembly and the start of the TBM determine complexity at this time. This is when TBM and soil interact for the first time; this is when the adequacy of the decisions prove to be true or wrong. The TBM is new to the crew and so are the soil conditions. In addition, the transition between concrete starting pit and soft soil is critical. Typically, there is a treated transition zone, where injections improve the soil quality. Another concern is the water tightness of the starting pit. We can note by this example that task complexity has different aspects. While the complexity of creating the TBM tool is now small, the complexity of using it is great. Altogether, task complexity remains high at this time.

6.4.2.2 Social Complexity
Starting a TBM drive is always challenging. Not everything will work as planned. There is a high pressure on the tunneling crew and the management. The owner and the sponsors will watch very diligently.

During planning and procurement, the social complexity arises primarily from the interaction between ICJV engineers, ICJV management, and TBM producer. During testing, the focus shifts from cognitive to operative activities with the involvement of the tunnel crew, ICJV management, sponsors, and owners.

Again, as one part of complexity diminishes, another starts to gain importance. The overall complexity remains rather similar; reductions are small.

6.4.2.3 Cultural Complexity
We find a shift in this domain. Cultural interaction now switches to the tunneling crew where local workers and foreign specialists work together. The tasks facing local workers might bewilder them. Many will never have seen a machine as complex as a TBM. Insecurity typically reinforces cultural divergence (history is full of examples!). However, arising complications are either on an individual or on a group level and easier to contain than problems on the ICJV level. Compared with social complexity, cultural complexity declines faster.

6.4.2.4 Cognitive Complexity
We can describe cognitive complexity at this time in the following way: learning continues and cognitive complexity evolves fast. In theory, the TBM should do a perfect job. In practice, this might not be true. Hence, we find a shift from theoretical to practical understanding. Fulfilling the construction contract includes perfect understanding in both areas. However, the practical understanding commences based on theoretical understanding and develops faster in comparison. The overall increase in cognitive complexity accelerates a little, but there remains a lot to learn.

6.4.2.5 Operative Complexity
The insecurity surrounding the first meters of tunneling invites comments from the outside of the ICJV by referring to experience from other projects. The tunneling crew is not

able to defend all actions since the success is still missing. Decision-making is still under outside influence. As this influence refers to other projects, it cannot be the best advice for the specific problems of the megaproject under consideration: operative complexity still develops slowly.

6.4.3 Mastering the Construction Technology

Testing finishes with the end of the learning curve. The tunneling crew has achieved planned progress. Attention shifts from learning to producing. The crew must master cycle times. It has reached a certain plateau of competence and must now anchor this level. Introducing new ideas at this time is counterproductive and disruptive.

Disrupting the mastering activity group = (obs.) An ICJV had to drive piles as deep foundations for bridge piers in a delta region. Top layers of cohesive soil with low bearing capacity characterize such areas. The engineers had the piles designed to be vertical. In this specific case, there were dispersed boulders in the top layers. When driving the piles, these boulders caused the piles to deviate from their vertical position. Very few of the piles were vertical according to design. Engineers had to redesign the pile groups, and very often, the pile-driving machine had to return to a pile group to drive additional piles. This delayed all consecutive activities on the critical path. It took more than a year to master the problem, a very long learning curve!

During this time, the home office came up with a new idea of inclined piles to save on quantities. A brilliant idea at a bad time! Had the project manager allowed the change in design, the chaos for pile driving would have been complete.

It is not smart to introduce new ideas during the mastering activity group. First, the crew must reach the required plateau and master the processes. Then, one can move on to the next level.

6.4.3.1 Task Complexity

Testing has already diminished the relevant task complexity: mastering reduces it further. Increases in task complexity are now due to external influences. In the case of TBMs or pile driving, this can be unforeseen soil conditions. Therefore, we have two overlapping tendencies: a considerable reduction of task complexity and smaller variations around this trend. Sometimes, it turns out that an ICJV cannot master a construction technology; this leads to dramatic increases of task complexity exceeding the initial level.

Increases of task complexity = (obs.): A contractor decided to use an overhead girder for the segmental construction of a bridge. Other contractors had used exactly this girder on two other projects successfully. While handling the overhead girder during the mastering activity group, it toppled off the bridge and killed three workers. The owner demanded an investigation, the contractor reevaluated the processes and construction restarted using the same overhead girder. Again, the handling proved to be precarious. The owner stopped the use of the overhead girder, and the contractor had to switch to another construction technology during project execution.

This increased project task (and social) complexity above the initial level.

6.4.3.2 Social Complexity

During the mastering period, social complexity typically decreases considerably. Not only have the players within the ICJV settled into a productive mode but also all stakeholders in the complete system. Especially, the interaction between owner and ICJV should concentrate on performing.

However, the ICJV has not completely mastered the processes, and everybody pays close attention to what happens. Any hiccup can cause disproportional increases in social complexity. If accidents happen, then the press will cover it and even politicians might become involved.

6.4.3.3 Cultural Complexity

Similar to social complexity, cultural complexity decreases if everything goes more or less according to the plan. However, any major problem might accentuate cultural differences. In such cases, culture is not the cause of problems but rather another area, which serves as a rallying point to express divergent approaches.

6.4.3.4 Cognitive Complexity

At no other period, our construction understanding of a megaproject grows faster. Cognitive complexity increases rapidly and provides the specific understanding required for the singular requirements of the focal megaproject. The project team develops the specific understanding alone. All other stakeholders not deeply involved fail to undergo the same cognitive development. This includes all supervisory levels above the ICJV and might include the owner.

Therefore, it becomes important to justify every step very carefully and with the necessary detailing. As many top managers (and sometimes owners) are not able to spend much time for the project, trust in the ICJV might replace the lack of deep understanding and we earn trust by signaling.

6.4.3.5 Operative Complexity

Coupled with the increases in cognitive complexity, operative complexity also develops considerably. It becomes hard for somebody outside the daily operations to give good specific advice, as this will come from experience in other projects. Singularity does not allow for an experience transfer in this period. The project management team becomes rather independent in their actions.

If trust is lacking for some reason such as a progress below expectation, then the operative freedom will be limited adding further hurdles to the progress. It can become a vicious circle.

6.4.4 Stabilization of all Construction Processes

The crews will have reached a new productivity level at the end of the mastering activity group. From here on, productivity will remain constant unless managers or engineers induce external changes or crews find autonomous improvements. Variations from the normal productivity level can be substantial during the mastering activity group. The task becomes to reduce such variations almost to zero during the stabilization activity group.

Figure 6.7 shows a solid line indicating the average productivity. The productivity increases (i.e. the input for a certain output decreases) in this case by 30% during the testing period. It remains constant during the mastering and stabilization activity group. The variations might be extreme during testing and considerable during the mastering activity group. They approach zero at the end of the stabilization activity group.

6.4.4.1 Task Complexity
The task complexity becomes small, and the focus shifts to eliminating the variations. There are internal and external reasons for variations. When building a high rise or a long bridge, external influences are not very important, the weather being a typical one. This is different for tunneling where the soil conditions are partly unknown and changing. It is much easier to achieve constant progress when soil is not a major factor.

6.4.4.2 Social and Cultural Complexity
In typical cases, social and cultural complexity are low by this point. All stakeholders should be in a performing mode. Compared with task complexity, both are higher. While we reduce

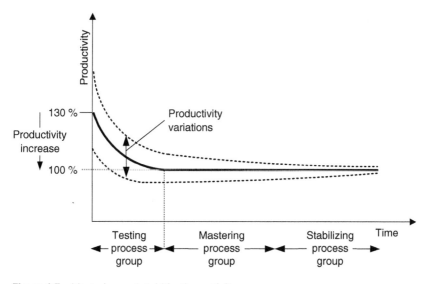

Figure 6.7 Mastering and stabilization activity groups.

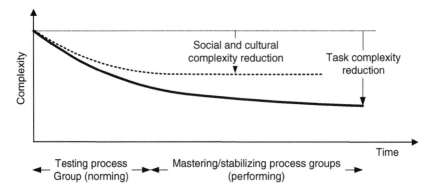

Figure 6.8 Reduction of task, social, and cultural complexities.

task complexity to zero at the end of the project, social and cultural complexity remain virulent reaching even beyond the end of the project.

Figure 6.8 shows a comparison of task and social/cultural complexity development during the mastering and stabilization activity groups. The social and cultural complexity will vary around the "normal" level from the end of the mastering activity group to the end of the project. As always, in human relationships, conflicts and periods of harmony will alternate. A complete breakdown in trust can lead to skyrocketing levels of social and cultural complexities. I have chosen the starting point of task complexity to be the same as for social/cultural complexity. This is not true in reality; however, we cannot measure the three complexities quantitatively on the same scale. Important is the difference in the development, and we know that social and cultural complexities cannot reach zero.

6.4.4.3 Cognitive and Operative Complexity

Cognitive and operative complexity during the stabilization activity group do not increase at a high speed. Both stabilize, which means that the project team gains confidence in their understanding of the megaproject and their actions. This increased confidence radiates to outside stakeholders such as sponsors and owners. The satisfactory progress that typically develops at this time is the concrete basis for the confidence.

Now, it is important not to interrupt the processes that are running smoothly. Larger changes or adjustments are not yet welcome.

6.4.5 Routine Processes

All megaprojects should have a long routine activity group. New demands arise for the project management team at this time. The hurly-burly period is definitely over, and the focus changes to maintaining progress and quality, grinding it out every day. This requires constant attention to similar processes. The task variety is much smaller.

Project manager = (obs.) Companies like to keep the same successful project manager in charge of a megaproject from the beginning to the end to ensure constant and detailed access to information. However, the question is whether the same project manager, who can deal with the pressure and constant changes of attention in the beginning, is also well suited to ensure constant routine quality. This certainly demands many different qualities from one person. I can satisfy my own interests best in the first half of a megaproject, and I feel that I lack the constant attention required for the second phase. I do not know a single project manager who is top with regard to both demands. Maybe it is better to find two managers who excel in those different phases.

The routine activity group provides the best opportunities for improvements. Seldom, these are dramatic changes to process or product. Very often, we find gradual changes or adjustments in the form of continuous improvement processes. Quality audits are typical examples for changes induced by engineers (top down). Widespread are also improvements by the crews. Gradual changes to the erection process of segments for a bridge accelerated the process from 2 days to 1 day. The 2-day cycle was already a world record. Doubling the speed is quite an achievement, and no engineer was involved.

6.4.5.1 Task Complexity

During this long period, the task complexity is at its lowest level before it becomes zero at the end. It is not zero now because of external influences or internal mistakes.

 Internal mistakes = (obs.) A project manager was in charge of a bridge project in Northern Africa. The day before concreting, the formwork of one abutment caught fire and burnt down completely.

He later became the project manager for a segmental bridge construction in Hong Kong. When the crew placed the last segment, it fell into the water and vanished forever. This mistake washed away the grand opening a few days later at the same time. None of this was the direct fault of the project manager; it certainly increased task complexity consequently.

6.4.5.2 Social and Cultural Complexity

Social and cognitive complexity are relatively low; disturbances remain always possible.

6.4.5.3 Cognitive and Operative Complexity

Both are still growing but slowly. As operative complexity always follows cognitive complexity, it also increases at this time. All this is true unless a continuous improvement process shows dramatic results. Then, cognitive and operative complexities increase with a leap.

6.4.6 Demobilization of the Project

Very often (and very unfortunately), "hectic" characterizes the end of a project. All of a sudden, many tasks need finishing and the handover date is fixed. This is not according to plan, but such deviations from plan occur on many sites. The task density increases due to additional resources, work processes are disturbed, and productivity slows down.

At the same time, the project management team must prepare the release of hundreds of managers and engineers and thousands of workers. Equipment worth many million USD needs selling. All this is a cause for an increase in complexity.

6.4.6.1 Task Complexity

Two or three strands of actions overlap during demobilization. First, planned construction activities end without slowing down much. Second, the demobilization activities start (releasing, selling, and documenting). Third, planned activities are very often late; nevertheless, all must end before the handover. More activities become concurrent with negative influences on each other.

6.4.6.2 Social and Cultural Complexities

The involvement of the owner increases toward the end, as he prepares to take over the project. Public interest also awakens again. If the pace accelerates at the end, then people become nervous and display a certain aggressiveness. This complicates relations.

Cultural complexity might also grow along with social complexity. Stakeholders might attribute problems to differences in national or business cultures (They never get anything finished properly!)

6.4.6.3 Cognitive Complexity

The project understanding still grows as loose ends connect and additional results become visible.

6.4.6.4 Operative Complexity

Due to the increase in cognitive complexity, operational freedom also grows. However, the renewed interest by the owner and sponsors might at the same time limit operative complexity.

6.4.7 Management Roles During Construction

Mintzberg (1973) promotes 10 different roles for managers and strongly stresses that manager must fill in more than one role at any one moment in time and switch between roles over time. There are quite a number of such role nomenclatures, and Mintzberg himself (2009) quotes a manager claiming that such descriptions of roles are lifeless while managing itself is full of life; in the eyes of this manager, such nomenclatures are useless.

I disagree with the manager quoted above, and I will intersect Mintzberg's management roles with the six activity groups. The resulting matrix tells me a story full of life. However, I first need to explain the different roles, which Mintzberg differentiates into three groups.

Interpersonal roles include the figurehead, the leader, and the liaison. A figurehead represents the ICJV at ceremonial events; these can be external (groundbreaking ceremony and opening ceremony) or internal (weddings and funerals). Leaders perform several tasks, among which are allocating work and information, along with supervising and motivating. The liaison maintains contacts outside the ICJV, especially with the sponsor and the owner; other partners in the case of megaprojects might be the government, authorities, police, and the public.

Informational roles include the monitor, the disseminator, and the spokesperson. The monitor scans the ICJV and the outside for relevant information. As a disseminator, he distributes this information to those who require it for their work. The spokesperson shares information with the outside, i.e. the joint venture board, the sponsors, the owner, and the public.

Decisional roles embrace the entrepreneur, the disturbance handler, the resource allocator, and the negotiator. The entrepreneur initiates action, and the disturbance handler provides immediate answers to unforeseen events. The resource allocator distributes resources, and the negotiator finds solutions for difficult problems that often require a compromise.

Of course, such roles are a generalization and a drastic simplification. They are easy to attack. However, I am at this moment not interested in an academic discussion of what constitutes management or how to describe a manager in detail. Certainly, the roles highlight the important activities of managers.

Table 6.2 shows my qualitative assessment of the importance of each role during one of the activity groups. The table does not show overlapping activity groups but distinct process phases; this is an additional simplification.

The role of a figurehead takes precedence at the beginning and end of a project. There might also be a ceremony in the middle of the project (over the hill). A leader is also required especially at the beginning and the end; leadership at the start is crucial. During the routine period, the ICJV should resemble a smooth-running machine; active leadership recedes into the background. Interaction with outside stakeholders is most important at the beginning. Building the team around the project is a prime task at the start; here, the relationship (liaison) building takes up considerable time.

Information is vital at the beginning and remains of some importance throughout the project; the monitor gathers this information, and the disseminator distributes it. Acting as a spokesperson is significant throughout; however, its interaction with team building makes it prominent in the early stages of the project.

All decisional roles in megaprojects are required mostly at the beginning; this is especially true for the entrepreneur and the disturbance handler.

With all the caveats that I have already mentioned, the table highlights one aspect clearly. The beginning of a megaproject profits best from someone with entrepreneurial skills and

Table 6.2 Management roles and activity groups.

	Planning/ procurement	Testing	Mastering	Stabilization	Routine	Demobilization
Figurehead	+++	++	+	o	o	+++
Leader	+++	+++	++	+	o	+
Liaison	+++	+++	++	+	o	+
Monitor	+++	+++	++	+	+	+
Disseminator	+++	+++	++	+	o	++
Spokesperson	+++	+++	++	+	o	+++
Entrepreneur	+++	+++	++	+	o	+
Disturbance Handler	+++	+++	++	+	+	++
Resource allocator	++	+++	+++	++	+	++
Negotiator	+++	+++	++	o	o	++

the middle phase from an administrator. I am aware that the connotations of the word "administrator" are partly negative. In the case of a megaproject, this discredits the corresponding activities without any reason.

6.4.8 The Course of Complexity throughout the Activity Groups

Figures 6.9–6.13 summarize the discussion of the seven activity groups in a graphic form. I have discussed each activity group in detail and choose as dimension for the *x*-axis not time but activity groups with equal spacing. Nobody should infer a time axis where there is none. Typically, the routine activity group will take up more than two-thirds of the project time. Spacing along a time axis would be very different.

I show all complexities to start at 100% on the *y*-axis. Thus, I normalize the real data. The five complexities have different qualities, e.g. 100% for task complexity is not the same as 100% for social complexity. It is impossible to determine quantitative values for complexity. We use judgmental expert opinion to assess complexity qualitatively. Therefore, only the course of complexity development is instructive.

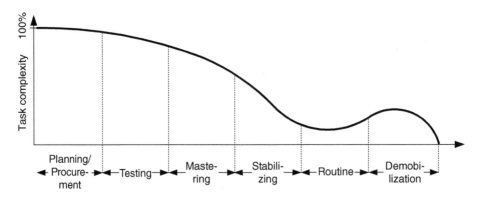

Figure 6.9 Course of task complexity.

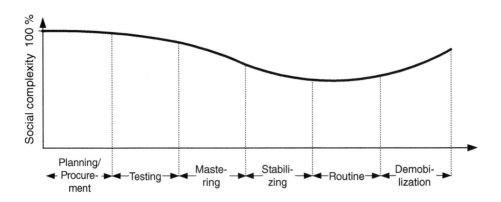

Figure 6.10 Course of social complexity.

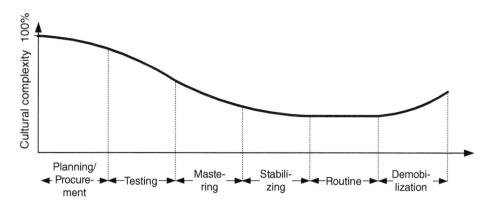

Figure 6.11 Course of cultural complexity.

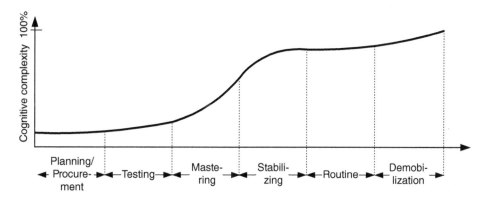

Figure 6.12 Course of cognitive complexity.

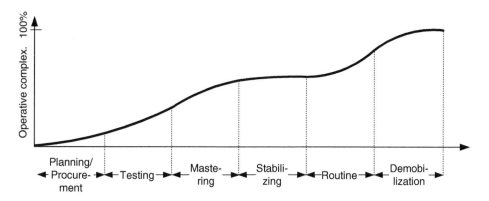

Figure 6.13 Course of operative complexity.

The courses of complexity shown refer to a well-managed project. Many events can occur that increase complexity. Thus, the courses approach the ideal case; this is, for many projects, typical; for others, it remains a guideline of what could be.

6.4.8.1 Task Complexity

Summarizing the discussions about task complexity over the activity groups, we know it decreases slowly and more sharply during the mastering and stabilizing stages. Then, it will fluctuate around a low level during the routine activity group and increase during the demobilizing stage.

6.4.8.2 Social Complexity

Social complexity remains virulent throughout the project and even beyond its limits. Once the project shows progress as planned, some stakeholders stop focusing their attention on the megaproject in question. Accordingly, social complexity decreases, only to increase again toward the end of the project.

6.4.8.3 Cultural Complexity

Cultural complexity diminishes faster than social complexity on the organizational level. Culture can cause problems, but often it serves as scapegoat for problems on the interpersonal level.

6.4.8.4 Cognitive Complexity

Cognitive complexity does not start at zero because the bidding team passes information and understanding of the situation to the project team. After a slow start when the overwhelming task complexity needs sorting out, cognitive complexity increases fast during the mastering and stabilization activity groups. Without improvements during the routine activity group, cognitive complexity increases slowly toward the end.

6.4.8.5 Operative Complexity

Operative complexity develops faster than cognitive complexity because the project team learns fast as compared to other stakeholders. The difference in knowledge and understanding provides for operational freedom. A second phase of a strong increase is during the routine activity group. Assuming everything progresses smoothly, nobody will place restrictions on the project team.

7

Descriptive Megaproject Management Model

The descriptive management model is the centerpiece of megaproject management. It is the cognitive megamap gathered from the experience of more than 50 managers on megaprojects. Not every project manager has the full model in mind, but few contest the validity of a category. I have gathered the data from observation and formal interviews. International project managers have checked the model and adopted it.

The main lesson of the model is that there is no single most important factor for megaproject management. On the contrary, managers must pay attention to and keep in mind a number of activities simultaneously. Depending on the situation, one activity comes to the foreground and another fades into the background. Managers switch from one activity to another at a fast pace. During the different stages of the project, different activities become more important. Due to delegation, some project managers focus on some activities while others in the top management team focus on other activities at the same time. From one project to the next, some activities are more relevant than others; the balance changes all the time. Unfortunately, I can give no advice to the balance for a specific project; nobody can do that. It is a search process of the top management team. They have to find a promising way forward as a team.

The model should be complete. I would like to lay claim to such a statement. However, doing so would deny the diversity of megaprojects and the importance of dynamic. We have to adjust our cognitive maps from time to time, and this is also true for the model as megamap.

The model allows for downscaling, i.e. we can use it for all types of projects. The smaller the project, the more activities become unnecessary or even counterproductive. The category "sensemaking" has absolutely no importance when building a one-family home. This is an asymmetry: downscaling is possible because we can just omit categories that we deem to be of no importance. Upscaling from a smaller model is more difficult, if not impossible, because we have to think about categories that are not clear.

Figure 7.1 shows the descriptive model and contains five main categories of tasks, functions, or dimensions. Referring back to Mintzberg's general management model, all these activities involve information (communicating, framing, scheduling, and controlling),

Advanced Construction Project Management: The Complexity of Megaprojects,
First Edition. Christian Brockmann.
© 2021 John Wiley & Sons Ltd. Published 2021 by John Wiley & Sons Ltd.

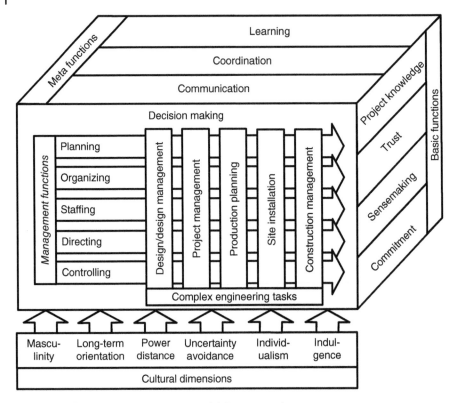

Figure 7.1 Descriptive management model for megaprojects.

people (linking and leading), and action (doing and dealing). The five categories with their fields of knowledge are:

- Complex engineering tasks (civil engineering)
- Management functions (business administration)
- Meta functions (business administration and psychology)
- Basic functions (sociology and psychology)
- Cultural dimensions (sociology)

This demonstrates that the required knowledge advances far beyond engineering. Looking at public infrastructure projects often demands additional knowledge of politics. More than a few project managers have resigned from their jobs because they did not like political interference in project operations. This happens to those working directly for the owner. Contractors are often shielded from carrying the burden of politics; they just have to bear the consequences, which might be hard enough. Project management companies working directly for the owner typically have to deal with political issues.

7.1 Management Functions

First, I will introduce the functions in some detail in this chapter. Then, I will tie the functions back to the discussion of project complexity in Section 7.2. This serves to build a bridge between the descriptive management model and complexity as well as systems theory. The abstract thoughts on complexity will become more practical.

7.1.1 Complex Engineering Tasks

The complex engineering tasks encompass design and design management, project management, production planning, site installation, and construction management. Most civil engineering programs at universities provide basic knowledge in these areas. The complexity of megaprojects signifies the need to develop this knowledge further, i.e. to advance construction project management.

7.1.1.1 Design/Design Management

Design refers to the product design, i.e. the design of the structure. In the case of design/bid/build contracts, the owner provides the design with the help of architects and engineers. If the owner chooses a general design consultant, this company becomes responsible for coordination. If he chooses separate design contracts, he must either manage the integration himself or designate a company to do so. In all these cases, design management is not the task of the contractor. This changes for design/build contracts; here, the contractor controls the design process. The required integration can include different subjects. Railway design must integrate infrastructure (civil engineering), electrical power, signaling and communication (electrical engineering), rolling stock (mechanical engineering), and human behavior. The responsible organization needs to master the management activities (planning, organizing, staffing, directing, and controlling) for design management. In megaprojects, design management becomes at least as demanding as the design itself.

7.1.1.2 Project Management

The Project Management Institute (PMI) codifies project management practice through the publication of the Project Management Body of Knowledge (PMBOK Guide 2017). The PMBOK identifies 10 knowledge areas: (i) integration management, (ii) scope management, (iii) schedule management, (iv) cost management, (v) quality management, (vi) resource management, (vii) communication management, (viii) risk management, (ix) procurement management, and (x) stakeholder management. I would like to add two more areas – (xi) contract management and (xii) health, safety, and environmental management – to complete the list. Again, knowledge in all these areas is basic project management knowledge. However, megaprojects add more complexity to the tasks.

7.1.1.3 Production Planning

Production planning is a term adopted from lean construction management. It describes a more detailed approach than conventional planning. Often, megaprojects require such detailed plans, even if the contractor does not apply lean principles. Production planning can demonstrate the difference between planning by the owner's project management team and by contractors. Contractors analyze every plan minutely, owners more generally. Of course, this comparison applies only for contents that both parties use in a similar way, such as scope, schedule, cost, and resources. The owner's and the contractor's stakeholder management certainly address partially different groups, so the claim of "more detail" cannot be upheld for this knowledge area. The same holds true for communication, risk, and procurement management.

To exemplify the idea, cost estimates provide a good example. Quantity surveyors or cost engineers preparing a budget for the owner need to rely on the historical data of cost elements such as walls, columns, doors, or windows. In case of a wall, they will work with a value of, for instance, USD/m^2. They are not well-prepared for shifts in price trends. Contractors get information directly from construction activities. They know the productivity of labor or equipment since they continually buy materials on the market. This information allows them to prepare an estimate of a wall based on labor, materials, equipment, and subcontracts.

7.1.1.4 Site Installation

Academics often treat site installation as a topic without much emphasis. This is wrong; the site installation provides the factory, in which we produce our product. Admittedly, the factory has no roof and the detail of planning is well short of a production line in the automobile industry. However, in megaprojects, the difference between site installation and production line can be small. When a project such as a tunnel requires precast elements in large numbers, robots and continuous workflows ensure at least partial automation.

7.1.1.5 Construction Management

Construction management is the daily business of assembling the structure with very large amounts of resources based on elaborate work methods. A project with a volume of 2 billion USD might require a workforce of 5000 in a developing economy. The number would be smaller (maybe 3000) in a developed economy. The main reason for this is typically higher productivity due to better training and more benevolent climatic conditions. Construction management in such cases means that 5000 workers must know where to start their work every morning and what to do, while location and work keep changing. Supervisors must know this a week in advance and coordinate the large workforce. The goal is not to waste time and to minimize mistakes.

7.1.2 Management Functions

Management functions include planning, organizing, staffing, directing, and controlling. This is very close to the POSDCORB concept developed by Gulick (1937). POSDCORB stands for planning, organizing, staffing, directing, coordinating, reporting, and budgeting. Koontz and O'Donnell (1955) introduced the classical five functions mentioned above. The functions suggest a neat sequence: make a plan; then create an organization with structure, schedule, and rules; move on to fill the structure with people; tell them what to do, and control whether the results are as planned. Unfortunately, this is not what happens on construction sites or elsewhere. A project manager who uses his first day on-site for planning will have lost the respect of all subordinates by the evening. The demands of the day dictate the sequence of the activities, and it is neither wise to neglect the demands nor have the demands drive the daily schedule exclusively. The day of a project manager will contain some or all of the management functions in any imaginable order; the day is not predictable.

7.1.2.1 Planning and Controlling

Planning and controlling are twin functions. With plans, we formulate goals; by controlling, we check goal achievement. Whenever there is a deviation from the plans, corrective actions become necessary. In a next step, preventive action should avoid making the same mistake again. The main tool is a comparison between planned and actual data. Plans in construction reach far into the future. Since nobody can foresee the future, plans must evidently be wrong at times. Therefore, we need to check whether a deviation is due to mistaken planning or execution.

7.1.2.2 Organizing and Staffing

Organizing and staffing are also at least partially twin functions. With an organization chart, we create a structure that needs staffing. Staffing means finding people that fit into a specific place in the organization chart. Of course, organizing also means creating a schedule and rules, and staffing means additionally administering and motivating people.

7.1.2.3 Directing

By directing, managers give instructions to the members of the organization. I prefer the term "directing" over "leading" to avoid any mystification of the leadership concept. As discussed, leadership is highly influenced by culture; "directing" seems much more innocent. We can achieve coordination in a project through directing as well as organizational rules.

All management functions require attention for a certain amount of time until the attention shifts to another task: they are discreet functions. This distinguishes them from meta-functions, which are continuously ongoing, although evidently not always in the foreground.

7.1.3 Meta-functions

Meta-functions comprise decision-making, communication, coordination, and learning. Watzlawick et al. (1967) explain their character by their first axiom of communication: "*You cannot not communicate.*" I think the same holds true for decision-making as we have to decide our next action or move continuously. Coordination goes on unceasingly in a megaproject, either by directing or by following the behavior of other people and by observing rules. Finally, learning never stops. All four activities require a lot of our brain processing activity at times and very little at others.

7.1.3.1 Decision-Making

Decision-making has received a lot of attention in management literature. The big question for megaprojects is how to make a decision without complete information or structure in a highly complex system. It is certainly not a time for rushing to conclusions or procrastination.

7.1.3.2 Communication

Communication is the basic tool for megaproject managers, and they prefer face-to-face communication. Communication follows hierarchy, informal relationships, and cultural fault lines. A lack of language skills often influences the last factor. All general management activities Mintzberg (2009) mentions (framing, scheduling, controlling, linking, leading, doing, and dealing) require communication for transmitting and receiving information, motivation, and coordination.

7.1.3.3 Coordination

Coordination is the largest problem in megaprojects. This statement refers to the coordination of information and action. Every day, thousands of workers and hundreds of employees arrive at the construction site. They are facing an ever-changing environment. Not very often can they do today what they did yesterday. The magnitude of tasks, people, and information demands constant coordination.

7.1.3.4 Learning

If we take singularity as the main characteristic of a megaproject, then learning becomes paramount. Singularity designates a new task, and therefore, nobody has the experience to master it without a good dose of very fast learning on the way.

7.1.4 Basic Functions

Basic functions are to a megaproject what oil is to an engine: without a good dose of basic functions, the megaproject will grind to a stop. It is to this group that project knowledge, trust, sensemaking, and commitment belong.

7.1.4.1 Project Knowledge

Project knowledge is not zero at signature. Sometimes, large teams have worked for more than a year to prepare the bid. Expenditures of 10 million USD are not exceptional. Typically, the bidding team hands over this knowledge to the project team during a kick-off meeting. Given the amount of knowledge amassed, the meeting might take several days. Even if the kick-off meeting is perfect, it will not be possible to transfer all available knowledge to the project team, which needs to get all available information as quickly as possible. All this requires sound knowledge management.

7.1.4.2 Trust

Trust facilitates many tasks and is indispensable in megaprojects to reduce complexity. The project team can only hope to master the massive workload by delegating work to its members. If the top manager insists on full control, the project will never take off. It requires more time than even the most diligent top manager can provide. Delegating large work packages requires trust among team members. This holds true even if the team members have never worked together before. Trust in megaprojects cannot grow with time; it is required from the start. It is easy to assess the problem when thinking about the phases of team building (forming, storming, norming, and performing). The team members must establish trust during the forming period, since trust will come under fire during the storming period. This form of trust is not the normal one, and I will call it "necessitated trust," which is very different from built-up trust. Built-up trust requires time, and it will replace necessitated trust during the project.

7.1.4.3 Sensemaking

Not many authors discuss sensemaking as part of project management. However, if we do not manage it, team members will set out to attribute their own sense to communication, decisions, and actions. This typically manifests itself as rumors spread by gossip. The project management team can counter rumors only through open and transparent communication management. They need to provide facts and a framework for understanding. The saying that everyone is entitled to his own opinion but not to his own facts has two implications for megaprojects. Firstly, people with knowledge must provide the facts. The project management team members – especially the project manager – have the largest amount of facts at their disposal. This implies that sensemaking is a task for top management. Secondly, project participants must accept facts and consider them for communication and action.

7.1.4.4 Commitment

Commitment describes how much project participants identify themselves with the megaproject and how hard they are willing to work for its benefit. In case of international construction joint ventures (ICJVs), mother companies send employees to work on the project. These employees know that they will return to their mother companies at the end of the project. Their careers are enfolding in those companies. Different partner companies in an ICJV seldom have the same goals. This poses a loyalty conflict among members of ICJVs; they have to choose between working for their own company or the ICJV. The megaproject can only achieve maximum success through employees and workers who are completely committed to the ICJV.

7.1.5 Cultural Dimensions

Hofstede has defined six dimensions to characterize national cultures and gathered data for a large number of countries to allow for comparisons. The dimensions are (i) power distance, (ii) uncertainty avoidance, (iii) individualism, (iv) masculinity, (v) long-term orientation, and (vi) indulgence. The six dimensions form a model of the construct called "national culture," and, as with most social models, the model invites criticism (McSweeney 2002). However, the model provides a good approach to cultural management of ICJVs.

7.1.5.1 Power Distance

Power distance describes the difference between powerful and less powerful members in a society. It depends on acceptance by the less powerful. The assumption is that members of a certain national culture carry this framing into the megaproject, i.e. one cannot forget one's cultural training. People with high power distance respect hierarchies, and people with small power distance prefer lean organizational structures.

7.1.5.2 Uncertainty Avoidance

Uncertainty avoidance defines the attitude toward the future. People with high uncertainty avoidance tend to worry about the future. In megaprojects, these people try to shape the future by planning; sometimes, they even confound a plan with the future. As plans are indispensable in megaprojects, a certain acceptance of uncertainty is beneficial: Too much planning will lead to rigid behavior.

7.1.5.3 Individualism

Individualism is one end of a continuum that has collectivism at its other end. Individualistic societies prefer reliance on oneself, while collectivistic societies prefer a tightly knit network of relationships. This influences the style of working and decision-making in megaprojects.

7.1.5.4 Masculinity

Masculinity also delineates a continuum, with femininity at the opposite end. Masculine societies are more oriented toward competition, and feminine ones toward consensus building. Both terms are descriptive and do not promote a preferred norm. It is easy to see how this dimension influences communication and decision-making.

7.1.5.5 Long-term Orientation

Long-term and short-term orientations are again opposites. Long-term orientation furthers pragmatic approaches to immediate problems, while short-term orientation leads to attitudes that are more rigid. The now is more important than tomorrow. People with a short-term orientation dislike changing plans and decisions, even if their implementation becomes arduous. Long-term orientation, on the other hand, allows for flexibility in handling today's problems as long as this does not jeopardize long-term goals.

7.1.5.6 Indulgence

Indulgent societies are those that allow enjoying life and having fun. Restraint societies, on the other hand, set strict social norms for behavior within the society. For each megaproject, the top management must first define beneficial norms and standards; at the beginning of a megaproject, flexibility is beneficial. Accordingly, project management will face problems when aiming for a restraint environment during the first phase of a megaproject. Projects in general provide for less restraint environment than companies do.

7.2 Management Functions and Complexity

In Chapter 4, I introduced the concept of complexity with five dimensions, i.e. task, social, cultural, cognitive, and operative complexity. In the following subchapters, I will explain how we can connect these dimensions and the different management functions of the descriptive megaproject management model. Most dimensions have several attributes, and it is, therefore, difficult to provide a bijective match. In most cases, the strongest characteristic will determine the match.

7.2.1 Management Functions and Task Complexity

Tasks in megaprojects tend to have very little structure in the beginning. Once the top management has started developing such structures, i.e. rules and norms, they are consolidated through repetitions; in the end, tasks become highly structured, sometimes to the point of being boring. The lack of structure does not allow for analyzing the task easily, and therefore, we need to tackle it in its full complexity. At the beginning, most tasks are highly challenging. A team effort is best-suited to solve complex tasks. Thus, task complexity and social complexity remain entangled.

Design is clearly task-oriented, design management less so. If we think of design management as a tool to complete the task of designing, then we can see both as task-related. The same is true of project management, with the difference that, with a design, we create a product with project management processes. Production planning provides the planning of construction activities, from schedules to budgets and method statements to resources. These plans then become tools. The site installation is the factory layout for the megaproject. Construction is the process of creating a physical object, i.e. the contracted structure (Figure 7.2).

Planning is also a process that is mostly task-oriented. The ensuing plan is a thing, even if a team might have elaborated it. We can reduce task complexity through the information contained in plans (schedule, investment plan, staffing plan, etc.). The same holds true when we create structures through organization charts, schedules, and rules. Staffing is more difficult to evaluate. On the one hand, we are looking for a match between the applicant and the bundle of tasks assigned to a position as well the qualifications required. On the other hand, the interviewing and evaluation of the applicant is a very social endeavor. Here, I prioritize the mechanistic view of matching and identify staffing as mainly solving a task. The discussion above also applies to controlling because it determines relationships

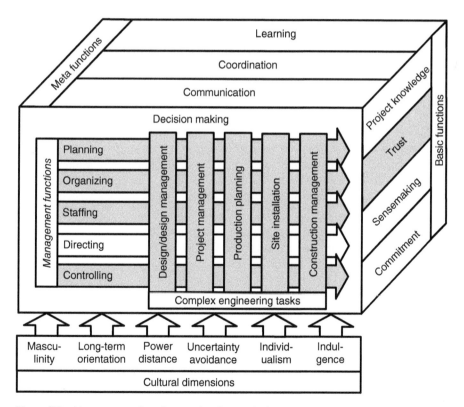

Figure 7.2 Management functions and task complexity.

between people. By controlling, we check whether we achieve the goals that were set out by plans. The applied tool is an analysis of planned versus actual (e.g. planned versus actual costs).

The most difficult definition is that of trust as a task. However, as I have explained already, trust is an indispensable tool at the beginning of a project. This is the most important aspect and permits defining trust as a task.

7.2.2 Management Functions and Social Complexity

Social complexity describes how multi-organizational and multi-personal a function is. In a megaproject, there will typically be several hundred employees and several thousand workers: it requires many people to solve the tasks. A stakeholder analysis will show how many different organizations are involved.

Directing is the function that helps to solve problems on a personal level. We need to direct people when there are no other means of coordination available. In the beginning of megaprojects, there are no plans, no processes, and no structures: this is the high time of directing. However, the function remains important throughout the project.

Communication is clearly a social function, as it depends more on relationships than just facts. We can achieve coordination in megaprojects by different means from directing to programs, i.e. from personal to technical coordination. Personal coordination is more important throughout, and it is central at the beginning (Figure 7.3).

7.2.3 Cultural Dimensions and Cultural Complexity

Culture provides megaprojects with the gift of diversity. In all cultures, people have developed norms and methods to solve problems they typically encounter. The more cultures there are, the more possibilities are available to tackle problems. This is the positive side of culture, and experienced global managers and engineers always pay attention to this aspect. The negative side are problems that arise from people of different mindsets working together without being aware that these differences might lead to misunderstandings. The increase in complexity due to different cultures also provides a variety of tools to solve problems.

In many megaprojects, we find more than 20 nationalities working together. Engineers around the world elaborate the design, and some of the procurement is global as well. Construction remains local. We can best solve this dichotomy with a combination of local and global personnel.

Power distance affects organizing, directing, controlling, decision-making, communication, and coordination. High power distance leads to authoritarian or patriarchic solutions, and low power distance to more egalitarian ones. Managers in an ICJV from countries with both (high and low power distance) must determine which path to follow.

Uncertainty avoidance relates especially to planning. The higher the uncertainty avoidance, the more pronounced is the tendency for thorough planning. The plans encompass design and design management, project management, construction management, site installation, and construction.

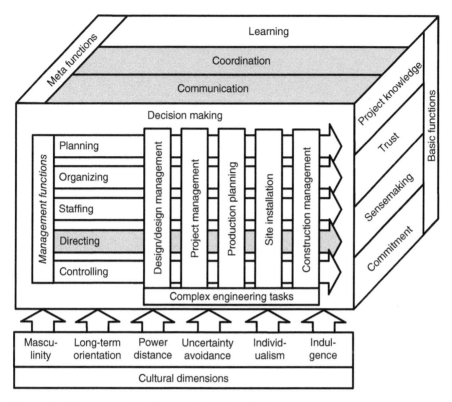

Figure 7.3 Management functions and social complexity.

Individualistic or collectivistic tendency determine relationships in a megaproject. This especially influences decision-making, communication, and coordination. A combination of both tendencies can be helpful as megaprojects demand individual leadership and consensus.

High complexity and a high workload leave us with no other choice than cooperation and joint efforts. As masculine cultures tend to be competitive, a feminine outlook is better-suited to solve complexity and deal with a high workload.

In megaprojects, we need both long- and short-term orientation. We must solve everyday problems, but one should never lose sight of the overall goals. Planning must always start from the end.

Megaprojects pass through different phases. In the beginning, indulgent behavior furthers innovation. After a while, restraint behavior guarantees a constant high-quality outlook. Again, diversity allows reducing complexity more efficiently (Figure 7.4).

7.2.4 Management Functions and Cognitive Complexity

Every megaproject is singular and requires a unique understanding. This understanding does not exist at the beginning; we need to develop it.

Learning is perhaps the most essential ability required from the top management team at the start of a project. However, before learning can take its place on the pedestal, we must

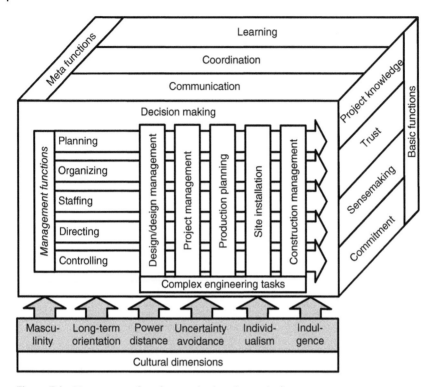

Figure 7.4 Management functions and cultural complexity.

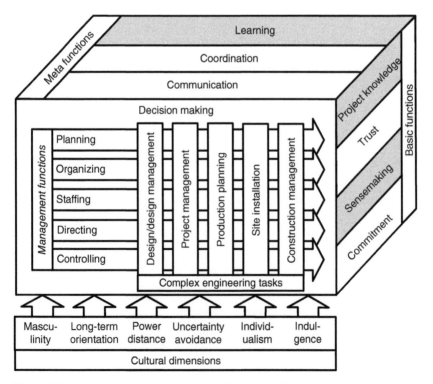

Figure 7.5 Management functions and cognitive complexity.

understand this importance. This is part of the overall sensemaking from one megaproject to the next. The faster the management team develops a cognitive understanding of the project through learning, increased project knowledge, and applicable sensemaking, the easier it is to reduce overall complexity.

The learning that develops the appropriate cognitive complexity (or mindset) of the management team allows reducing task and social and cultural complexities. It is a tool that can be used to break down these complexities (Figure 7.5).

7.2.5 Management Functions and Operative Complexity

We need not only a unique mindset for each megaproject but also unique actions. It is necessary that the management team is willing and able to develop their own approaches to solve problems through both decision-making and commitment.

As in many cases before, we reduce complexity by first increasing it and thereby generating tools for the reduction of complexity. In the short run, it would be more efficient to use tried and tested approaches that are available from previous projects. In the long run, this will lead to suboptimal results. As project complexity is often at its peak in the beginning, this unfortunately adds more complexity in a critical phase (Figure 7.6).

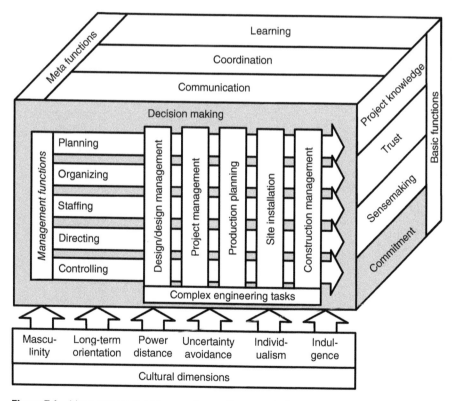

Figure 7.6 Management functions and operative complexity.

7.3   Combining Management and Complexity

In Section 3.2, I introduced Mintzberg's general management model with three planes (information, people, and action). In Section 4.2, I touched upon the complexity of megaprojects with five types of complexities, and in this chapter, I discussed the descriptive megaproject management model with 18 management functions and six cultural dimensions. Table 7.1 illustrates how these components can be tied together.

Table 7.1   Relationships between management and complexity.

	Management planes	Type of complexity
Design/design management	Information	Task complexity
Project management	Information	Task complexity
Production planning	Information	Task complexity
Site installation	Information	Task complexity
Construction	Action	Task complexity
Planning	Information	Task complexity
Organizing	Information	Task complexity
Staffing	Action	Task complexity
Directing	People	Social complexity
Controlling	Information	Task complexity
Decision-making	Action	Operative complexity
Communication	Information	Social complexity
Coordination	People	Social complexity
Learning	Information	Cognitive complexity
Project knowledge	Information	Cognitive complexity
Trust	People	Task complexity
Sensemaking	Information	Cognitive complexity
Commitment	Action	Operative complexity
Power distance	People	Cultural complexity
Uncertainty avoidance	People	Cultural complexity
Individualism	People	Cultural complexity
Masculinity	People	Cultural complexity
Long-term orientation	People	Cultural complexity
Indulgence	People	Cultural complexity

The table provides an overview of important tasks in a megaproject to which the project management team must pay attention. At different times, a different focus is required, so the attention must change from one topic to another. It also gives information about the type of approach which helps to deal with a certain topic (action, information, and people). Finally, it addresses the arena of complexity be it task, social, cultural, cognitive or operational complexity. On an abstract level, it can guide reflection and action.

8

Engineering Management

Engineering management combines aspects of engineering and management. As a result, it focuses on products of engineering as well as processes of management. The term contains instrumental, functional, and institutional connotations. We can say: (i) "Show me the schedule," i.e. the schedule serves as a tool; this is an instrumental engineering aspect. (ii) "Develop the schedule," i.e. the schedule is a task; this is a functional management aspect. (iii) "Check with scheduling," i.e. go to the scheduling department; this is an institutional management aspect.

As already briefly explained, the descriptive megaproject management model (Figure 8.1) contains five engineering tasks (together with the corresponding management functions and institutions):

- Design and design management
- Project management
- Production planning
- Site installation
- Construction

To simplify matters, I will stress the instrumental aspect when addressing one of these five tasks. This is arbitrary; at different times in the project, one or another aspect might take precedence.

Table 8.1 shows examples of the three different aspects for each engineering task.

8.1 Design and Design Management

We can differentiate between the process of designing a structure (functional view) and the design of the structure (instrumental view). For a book on construction project management, the process is more interesting. Typically, an international construction joint venture (ICJV) will hire one or more design firms. Its main task is design coordination.

8.1.1 Design Management

For many projects, we produce the designs in stages. Often, this includes an architectural design (form and function), a civil engineering design (stability and usability), and an

Advanced Construction Project Management: The Complexity of Megaprojects,
First Edition. Christian Brockmann.
© 2021 John Wiley & Sons Ltd. Published 2021 by John Wiley & Sons Ltd.

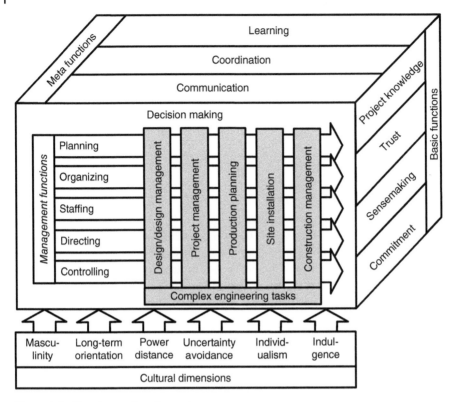

Figure 8.1 Complex engineering tasks.

Table 8.1 Aspects of engineering tasks and functions.

	Instrument	Function	Institution
Design	Calculation and drawing	Designing	Design department
Project management	Plans and meetings	Managing	PM office
Production planning	Plans and method statements	Planning	Engineering department
Site installation	Layout and logistics	Planning	Engineering department
Construction	Construction management plan	Managing	Site office

electrical/mechanical design (function and usability). This holds true for buildings. In civil engineering structures such as bridges, tunnels, or dams, an architectural design might be missing. There are exceptions to this rule, e.g. the Pont de Millau by Foster + Partners (Figure 8.2).

Traditionally, architects and engineers develop the design separately, with the architect in charge of coordination. Building Information Modeling (BIM) allows all parties to work concurrently on one project model. As megaprojects tend to be singular, they must be innovative, and one part of this is design innovations.

Engineers develop a design over time, and this requires design management. The process is divided into phases, and it allows for quality checks and coordination at the end of each phase. The quality checks might be internal or external with the help of independent checking engineers. Coordination also has internal and external aspects. Internally, we need to check for design clashes among architectural, civil, and E/M design. Externally, we must ensure that the design meets the expectations of the owner. I have previously called this alignment gates (Section 6.2). It is simply not economical to develop a full design and have it rejected by the owner. It makes much more sense to develop the design to a certain maturity level and get approval from the owner before moving to the next maturity level.

International construction often follows the US-American approach, with three to five levels. The original three levels of conceptual, preliminary, and detailed design often prove too crude; therefore, detailed design spreads into three levels: 35% detailed design, 65% detailed design, and 100% detailed design. The 35% detailed design serves one last time for coordination with the owner, the 65% detailed design for coordination with the independent checking engineer, and the 100% detailed design for coordination with the contractor.

The design is more or less a sequential process. As the beginning of the next phase depends on approval of the previous one, it makes no sense to work concurrently on two phases. In addition to the five phases shown in Figure 6.3, contractors must produce as-built drawings. All drawings up to the 100% detailed design are plans; the as-built drawings show the actual construction output. The differences should not be great, but there will be some.

It is necessary to manage the design process closely, especially for megaprojects. The design requires a large number of engineers (and architects if the structure is a building). Someone must coordinate the different engineering trades. Often, the design teams are scattered around the globe, working in different time zones. Distances between design teams complicate the development of trust and exchange of information. They make it impossible to walk to the colleague next door for a professional chat.

Figure 8.2 Pont de Millau by Foster + Partners. Source: SchrijverijDrenthe/Pixabay.

Design cooperation = (obs.) Design and construction cultures differ across the globe. Large European construction companies have a tradition of design/build approaches and, for this reason, have large technical departments. I started my career as a structural engineer in a subsidiary of a large company in Hamburg, Germany, which employed 80 structural engineers. In the head-quarters in Munich, the company employed 300 structural engineers. These are large numbers, and therefore, structural design expertise was extensive. The company would get closely involved in the construction aspects of 100% detailed design, and the firms responsible for this detailed design were aware of this behavior. US-American design firms have a tradition of providing the full design to a contractor who is happy to receive it without much comment. Pairing a European contractor with an US-American design firm will lead to misunderstanding, as the contractor will provide design input and the design firm will feel unduly criticized.

The quantity of drawings in megaprojects is substantial and can easily reach 10 000 in number. Calculations and communication accompany these drawings. The amount of information makes it necessary to use a document management system. Such systems depend on the documentation process, and this process needs definition. A documentation management plan defines a filing system for storage and a coding system for retrieval. The plan also contains rights (who can store, retrieve, and manage) and responsibilities (who must take what action). Cross-functional flowcharts illustrate this process and act as the blueprints for tailoring documentation management software to the specific purposes of the megaproject in question. This takes on another dimension when using BIM, as this approach advances the integration of the design process. In this case, there is need for the position of a BIM manager.

The documentation management can include a schedule management. In any case, a separate design schedule becomes necessary. The construction schedule is the basis for establishing such a design schedule. The construction schedule tells us the time at which the construction of a part of the structure will commence. From here, we can work back-wards by first determining the time required to prepare construction. Then, we need to consider the external review process. This is not easy because it is never clear how many resubmittals of drawings an independent checking engineer might request. This depends on the quality of internal quality management (QM) reviews, which again demand time. The duration of the design depends on the manpower available (Figure 8.3).

In the example in Figure 8.3, the design must start 12 weeks before construction. The design firm can control 4 weeks for the design and 2 weeks for internal reviews. This is not true for the external review process. Here, it is important to note that the contract between designer and owner determines the duration of review periods for the external checking engineer (who has a contract with the owner in this case).

The designer must define how to organize the internal quality management reviews. On the one hand, bad quality will affect the reputation of the design firm or even lead to claims against the firm. On the other hand, good quality management costs money. It is up to the design firm to strike an optimal balance.

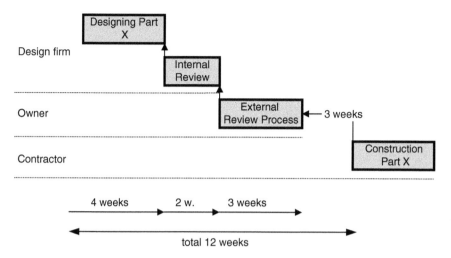

Figure 8.3 Example of a design schedule.

In many megaprojects, owners demand external reviews by an independent checking engineer. It is a decision that depends on the intentions of the owner. He will have to contract the independent checking engineer. As mentioned, it is imperative to fix the duration of review periods in the contract between the owner and designer (maybe 2 weeks for the first submittal of a design package and 1 week for a resubmittal).

It is also possible that the contractor wants to review the design to check for constructability before submitting it to the independent checking engineer. This adds a step to the process and requires additional time. It is part of the internal review process, which then has two levels. The first is the internal review by the design firm, and the second is internally by the ICJV.

8.1.2 Design

If it is true that megaprojects are singular, with a high degree of complexity, difficulty, and novelty, along with the requirement of specific resources, then we need to develop a design with unique answers to the challenges. We also have the opportunity because the large investment allows for an equivalent input to design. Often, design firms estimate the cost as a percentage of the project volume. Larger projects need more design hours, i.e. larger budgets, but the relationship is sublinear. The difference between a linear budget line (for normal projects) and the sublinear line for basic design input for a megaproject provides extra hours for innovation, and this innovation is available and necessary for megaprojects (Figure 8.4).

Satisficing (Simon 1955) is the heuristic used to finalize the design (i.e. determining the end of looking for changes or new solutions to the design). Satisficing means meeting an acceptable standard. The design firm can set this standard, or it can be the owner pushing the designer to further efforts. In the case of design/build contracts, it can also be the contractor.

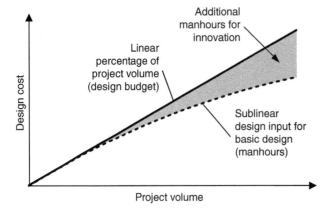

Figure 8.4 Additional manhours for innovation in design budget.

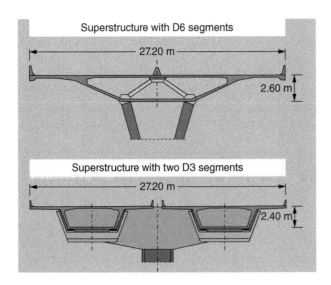

Satisficing as a heuristic = (obs.) Figure 8.5 shows two design solutions for an elevated expressway project. The solution at the bottom followed previous designs closely (superstructure with two D3 segments). The design/build contractor thought a better solution could be possible; the first design did not meet his internal level of acceptance. A second design firm received a contract to advance the design. Figure 8.5 shows this result in the upper part (superstructure with D6 segments). This design was radically new and saved large quantities (post-tensioning, rebars, and concrete) for the superstructure, the columns, and the foundations. It also facilitated construction.

Figure 8.5 Design options for an elevated expressway.

8.2 Project Management

The Project Management Institute (PMI) has codified project management knowledge and the Project Management Body of Knowledge (PMBOK) is widely used for megaprojects. However, the owner and the contractor are free to choose their project management approach and the corresponding methods and tools.

The PMBOK distinguishes five process groups: (i) initiating, (ii) planning, (iii) executing, (iv) monitoring/controlling, and (v) closing. These activities overlap and do not form a sequential process. It is for this reason that the PMI calls them "process groups" instead of "phases" (Figure 8.6).

The PMBOK concentrates on technical approaches to management. Relational or cognitive aspects are not in the foreground. I have included these in other parts of the descriptive megaproject management model (e.g. communication, trust, control, learning, project knowledge, and sensemaking).

With regard to processes, we can combine the Deming circle with the knowledge areas. The Deming circle includes four activities (plan, do, check, and act) and a methodology of continuous improvement or kaizen (Figure 8.7). In all knowledge areas, we must first have a plan, then implement it. Checking or controlling the results follows; we do this

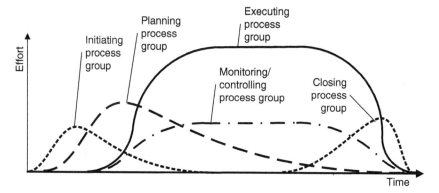

Figure 8.6 Project management process groups.

Figure 8.7 Deming circle.

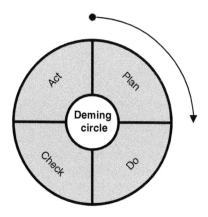

with the help of a comparison between planned and actual data. Acting means performing a root cause analysis in case of deviations between planned and actual data. A deviation requires corrective action; this means either correcting the error (actual) or the planning. People with high uncertainty avoidance (and accordingly, strong belief in planning) tend to blend out planning mistakes. This is neither logical nor practical. A deviation also requires preventive action; this means adjusting planning or implementation. Analysis (check) and corrective as well as preventive actions (act) describe the continuous improvement process. Plans and implementation (do) are inputs for the process. Through continuous improvement processes, we start a learning cycle with the goal of never repeating the same mistakes.

Very often, authors of project management describe a magic triangle (cost, time, and quality) or triple constraints (scope, time, and cost; Greiman 2013). There is absolutely no magic in such a triangle, and we cannot limit constraints to three factors. Should, for example, health, safety, and environmental (HSE) management not be part of the "magic"? My point is that it is not sufficient to focus on three constraints or even 10 knowledge areas of the PMBOK. The descriptive megaproject management model contains many more tasks, functions, and dimensions than a magic triangle or the knowledge areas of the PMBOK. Concentrating on triple constraints or 10 knowledge areas alone will almost certainly and without any magic lead to disaster.

I will discuss the 12 knowledge areas beyond the descriptions of the PMBOK in the following subsections. Those who are not familiar with the PMBOK should consult it concurrently.

8.2.1 Integration Management

The PMBOK describes how to develop a project charter and a project management plan.

Megaprojects might also require a business plan or feasibility study to show the chances and risks of the project and to procure financing. It will often take the form of a SWOT Analysis that opposes the external environment (opportunities and threats) and the internal environment (strength and weaknesses). A governance plan defines the values, norms, and rules guiding the megaproject. In international business, rules of compliance are of high importance for some countries. "Compliance" refers specifically to dealing with corruption. Corruption is an unpleasant fact in many countries, and construction is one of the industries most prone to corruption (www.transparency.org). The phenomenon exists in all countries; however, there are large differences among countries. As traditions vary, so does indulgence for corruption. While very few countries do not condemn corruption publicly (you have to do it to get international financing!), many countries accept corruption to some degree.

Plans for megaprojects must describe goals for corporate affairs (business and governance plan), legal affairs, project management, design management, construction management, and commercial and administrative management. A project management plan concentrates on project management. A project execution plan (PEP) coordinates all other plans in megaprojects. A PEP lists all other plans and describes them by defining the principal purpose of each plan by outlining its objectives, structure, processes, interdependence with other plans, responsibilities, and pertaining documents.

8.2.2 Scope Management

Scope is the sum of all products, services, and results. While it is complicated to define all these components, managing the scope development process is often inextricably complicated. With a bit of humility, we have to acknowledge that it is impossible to develop a complete and consistent scope for megaprojects in the beginning. There will always be changes. Therefore, we need procedures for managing scope changes; we specify such processes in the scope management plan.

Scope changes cause ripple effects throughout. Time and cost are especially affected. Accordingly, we should try to minimize them. One approach for the case of design/bid/build contracts is a design freeze before commencing construction.

Another way is discipline from the owner during construction, as he will either initiate or approve scope changes. Scope creep is a very dangerous phenomenon, where small changes are suggested and accepted every day. What is nothing for a day might amount to a lot after a year.

Scope changes are not necessarily bad. The Elbphilharmonie (Figure 1.2) is one example. The architects made changes to the project well into the construction period. The cost increased 10-fold, it took three times the planned construction time to finish, and the controversies were myriad. Yet, the result is proof that the scope changes were worth the money and the effort. The Sydney Opera House is a similar example.

Sometimes, it pays to rush into a project without a defined scope. This will increase construction cost and time while otherwise extending the operation time of the structure due to shortening of the overall time from first conception to start of operations. This additional operation time might generate money exceeding the amount spent for scope changes and construction time extension. Owners will have to consider the sum of capital and operational expenditures versus operational income.

8.2.3 Time Management

All schedules are wrong. An engineer who creates a correct schedule must be able to foretell the future over 3, 5, or 10 years, and such engineers do not exist. A bit of modesty tells us that all schedules are wrong. There are tools (schedule management software) and methods to approach future reality, and we must put these to good use. The basic method consists of creating many activities and determining the duration of each activity in an unbiased way (neither pessimistic nor optimistic). The underlying theory behind this method is the law of large numbers. This theory postulates that results from a large set of data will tend to approach the expected value while individual results might differ considerably from the expected value. A normal distribution (Gauss function and bell curve) describes the dataset with the expected value as mean. The complexity of megaprojects makes it easy to create a large number of activities; 10 000 activities might not be sufficient to plan and control a megaproject. It is also important that the activities share the same level of detail; otherwise, the influence of some activities might lead to a one-sided distribution. It is for this reason that we typically create schedules of different degrees of detail. Another problem that brings about one-sided distributions is the introduction of buffers for unforeseen events: they distort the outcome. Buffers are recommendable for schedules of little detail; they make no sense for construction schedules. Only the chosen duration of the activities should reflect the risk appetite for time. A contractor with a small risk appetite will plan longer durations for the activities, and one with a larger risk appetite will plan shorter ones.

Most scheduling software packages allow attaching resources to the activities. The resources can be quantities, manhours, or cost. All of these vary with time, and to check for the impact of schedule changes, an approach integrating resources is indispensable.

8.2.4 Cost Management

All budgets are wrong. This is true for the same reasons given for scheduling: nobody can foresee future expenditures. We can only estimate these. Milgrom (1989) puts this into an elegant formula: $X_i = C + \varepsilon_i$. Milgrom used auction theory to derive the formula. Most owners request two envelope bids. All interested contractors (open bidding) or all selected contractors (selective bidding) must provide two envelopes, one with the description of the technical approach, another with the estimated cost. This is a specific form of an auction. The interpretation of auctions often relies on two premises: (i) private values assumption, i.e. nobody knows his private costs; and (ii) common values assumption, i.e. all bidders in an auction face the same costs. Thus, X_i is the individual bid of a contractor in US dollars. C describes the cost of the project at its end based on the common values assumption; these are similar for all bidders. ε_i denotes the individual estimating error; this is the difference between the cost at the beginning of the project (X_i) and the cost at the end (C). We derive this estimating error from the private values assumption.

If all bidders show no bias in estimating, then the mean value is the only one without estimating error ($X_m = C$). The owner, however, will award the contract to the low bidder. In auction theory, we call this the "winner's curse." It is important to understand the implication: due to the bidding process, the winning estimate is wrong on the low side. The big question is whether the assumed profit exceeds the error.

To minimize the estimating error ε_i, engineers again use the law of large numbers. It is of paramount importance to break the project down into as many cost items as possible. The owner pre-structures the direct costs by his bill of quantities. The estimator must find his own breakdown for the indirect costs.

The way owners calculate costs (eventually with the help of quantity surveyors or cost engineers) is quite different from the way contractors draw up an estimate. This difference is due to the information available for both sides. The owner has no information about productivity or actual wages (labor cost), actual material costs based on signed purchase contracts, actual equipment costs depending on productivity, or subcontractor costs as specified through subcontracts. Contractors structure their estimates more finely and make better use of the law of large numbers. However, this does not safeguard contractors against estimating errors. The problem of megaprojects lies in the complexity of the estimate and dynamics that often change the assumption. In design/build contracts, every design change leads to quantity and quality changes that estimators need to consider. With several thousand cost items, one might easily lose control.

This means that we control costs against a benchmark that is unreliable, i.e. the error-prone estimate. Deviations might occur because of wrong planning or execution.

The bidding team will have a tendency to find fault with the execution and the project team with the estimate.

Psychology plays a role in estimating as well as in controlling. If there is pressure on the bidding team to submit a low bid, this will lead to a bias toward low unit prices that might not cover costs. If there is the chance to get a better than predicted profit during execution, the project team will be tempted to spend more money than necessary and be content to reach the estimated profit. It will get praise enough.

 Estimating bias = (obs.) Two companies joined to submit a bid. They both had experience with the tendered structure; however, they employed different construction technologies. They formed two estimating teams, each headed by a strong personality. When comparing their first estimates, the company with the higher bid adjusted their estimate by including optimistic assumptions of the other company. In the next round, the companies exchanged their roles. The two teams found themselves in a deadly downward spiral; the estimate in the end contained only optimistic assumptions. The result was a price half the cost of the previous projects. Both teams knew that the price level was dangerously low and still signed the billion-dollar contract. Egoistic competition led to risking the companies' survival.

Labor productivity and labor costs vary across the globe. The productivity of skilled workers in favorable environments is two or three times better than that of unskilled workers in unfavorable ones. The hourly labor cost in Australia (developed economy) is 60 times higher than in Vietnam (economy in transition). As a result, the labor costs for buildings in developed economies make up roughly 60% of the total costs, while in economies in transition, these make up only 10%. Inversely, the material costs in developed economies contribute 25% to total costs, and in economies in transition, this is 75%. In developed economies, we must focus on labor, and in economies in transition, on material handling. Productivity is of much greater importance in developed economies with regard to cost and time. In developing countries, productivity is important only for time.

8.2.5 Quality Management

Quality management has two aspects: quality assurance (QA) and quality control (QC). The owner defines the required quality in the contract and it is the contractors' task to specify procedures to achieve the contractual quality in a quality management handbook. QA plans quality procedures and controls these through audits. All procedures that prove insufficient will undergo a continuous improvement process (determination of the mistake, root cause analysis, and corrective and preventive action). With QC, we check physical results on-site.

The owner as well as the contractor conducts QC. In large megaprojects, there might be several hundred engineers working in QC for the owner. In order to be successful, we need clear quality definitions, clearly defined ways to measure the quality, and clear processes to deal with deviations.

8.2.6 Human Resource Management

If we assume the contractor's labor force for a megaproject to be 5000 workers, then we can also assume 10% of this to be the approximate number of employees (500). In case of an ICJV, the workers and employees come from different places (Figure 8.8). Typically, the local partner provides the labor force. The markets in some countries do not yield enough qualified workers, and then, the ICJV needs to contract foreign workers. This depends, of course, on the labor laws of the country where the site is located. The foreign partner will second either top management personnel (managers and engineers) or highly specialized operative personnel (supervisors and foremen). These are often much more expensive then local personnel. Only a high level of qualifications justifies such expenses, and these qualifications must be lacking in the local employment market. The local partner will employ the bulk of the top and operative management. Third-country nationals are another source for staffing. These are again highly qualified and expensive people.

The first step in HR management is the determination of the number of employees and workers. A first possible approach to achieve this is a bottom-up analysis: (i) Create a work breakdown structure with small work packages at the lowest level, (ii) assign a single package or several work packages to a hierarchical position, and (iii) build the organization structure from the lowest level by combining positions to teams or departments to the top level of the responsible project manager. A second approach is top-down: (i) Create an organization chart, and (ii) determine the number of positions. A third approach is a parametric: Here, we deduct the number of employees and workers from similar projects. In all cases, we can compare the results with the data from the estimate. This provides information about total manhours (workers) and employees. By loading the manhours into a schedule as resources attached to activities, it is possible to create a manpower curve in the form of a histogram. Unfortunately, all three approaches as well as the estimate are vague and mostly

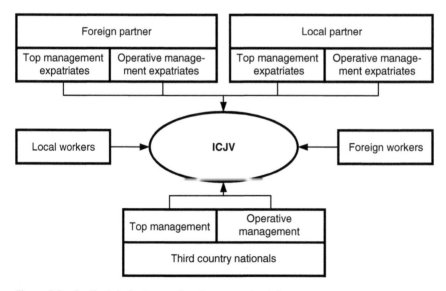

Figure 8.8 Staff origin for international construction joint ventures.

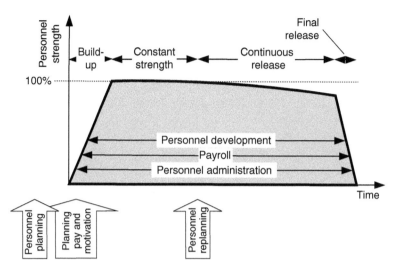

Figure 8.9 HR management activities.

provide inconsistent results. Therefore, the typical approach is more pragmatic: Most ICJVs develop the organization chart gradually, building it up from a core team at the beginning to full size. The information from the estimate serves as a guideline. As always, the estimate might be wrong, i.e. more or fewer workers or employees might be required.

Figure 8.9 illustrates the form of the personnel curve. In the beginning, there is the build-up to full strength. Then, the full strength remains constant. Somewhere in the middle of the project, a gradual release sets in. This happens in bridge projects when all foundations or all piers are finished. Due to short construction periods, the final release is rather abrupt.

Personnel planning starts before construction and is then continued. Sometime later, re-planning becomes necessary, i.e. adaptations of the original plans to actual requirements. Re-planning might sometimes focus on quality (hiring and firing) and sometimes on quantity (additional build-up or release). The shape of the personnel curve is, to a small degree, dynamic.

Planning pay and motivation forms a part of developing labor and staff contracts. To reach 100%, the administration department of an ICJV might have to negotiate and sign 5 000 labor and 500 staff contracts (given the numbers assumed at the beginning of this chapter).

These contracts require administration and payroll during the entire construction period. The responsibility for personnel development in an ICJV rests not only with administration but relies rather strongly on the activities of managers.

8.2.7 Communication Management

Communication requires planning and decisions with regard to communication technology, communication methods, and reporting. The first step, however, is a stakeholder analysis to determine whom to communicate with. It is also indispensable to check the contract for communication requirements. Often, contracts specify reporting formats at least partially.

Information and communication technologies (ICT) develop at a rapid pace. Twenty years ago, it was not possible to send a drawing as an email attachment – because of its size on the one hand, and transmission speed and mailbox size on the other. Today, this is not a problem. High-capacity laptops, smartphones, high-speed Internet and intranet, email, group communication, social media, virtual conferencing, and radio frequency IDs are just the beginning of digitalization. Our problem is not a lack of communication technology; it is the intelligent use of the technology.

Technical communication = (quote): "You cannot fax a handshake." Advertisement by Xerox, formerly one of the prime producers of fax machines.

Sometimes, incorrect use of ICT creates more problems than benefits.

Email = (obs.) The project manager of a multi-billion-dollar megaproject received approximately 150 emails per day. It took 5 minutes on average to answer one. The time varied from a click for spam and 30 minutes for a more substantial email. If we add this up, it takes 12 hours for answering the emails of 1 day. Accordingly, the project manager keyed answers into his smartphone during meetings and lost control of the meetings. In addition, he never controlled the project. The phone was smart, the project manager not.

The very difficult key point in communication is to give everyone all necessary information but not more, i.e. unnecessary information. Necessary information includes not only all the information required to perform a task or solve a problem, but also all the information required for orientation. This means all information required to meet the general goal toward which the task or the solution of the problem contributes.

If we are just looking at the possible communication channels of the presumed 500 employees (E), then we face the following amount of channels (C): $C = E * (E - 1)/2 = 124\,750$. This is impossible to manage directly. Only a communication hierarchy can help. An intranet solution with different access levels is not precise enough.

The project management team must formally report to the owner and internally to the joint venture board. The contents of the reports must be clear between the parties. A monthly progress report for a megaproject contains a large amount of data. One engineer might need 2 weeks just to write the report. Reporting is the tip of the iceberg. Below the waterline, there are systems to collect progress (scope), schedule, cost, and quality data as well as a discussion of problems and photo documentation.

8.2.8 Risk Management

Risk management consists of planning, identifying, analyzing, managing, as well as monitoring and controlling risks. We define risks as positive and negative future events for which

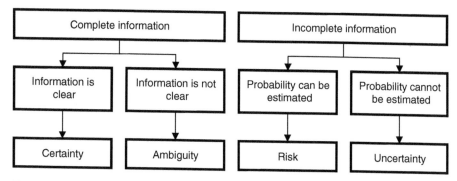

Figure 8.10 Information and resulting situations.

we can estimate a probability of occurrence. Risky situations describe a certain stage of information; others are certainty, uncertainty, or ambiguity (Figure 8.10).

With complete information (which we seldom have during the formative stages of a megaproject), we can either understand or not understand the information. In the first case, we face certainty and, in the second, ambiguity, as we do not know how to evaluate the situation. Incomplete information is typical for megaprojects. Sometimes, this incomplete information allows estimating a probability of occurrence; this describes risks. If we do not have enough information and we cannot determine a probability, then we are looking at uncertainty. These distinctions allow for a better understanding of the term "risk."

A risk management plan contains the following elements: (i) risk policy, (ii) risk scope, (iii) risk governance, (iv) risk management process, (v) risk communication, and (vi) awareness and training. The risk policy should describe a proactive approach through which we can identify risks and possible answers beforehand. The overall responsibility rests with the project manager. Nevertheless, risk management remains an integrative team effort with responsibilities to identify, report, integrate, and again, communicate risks. The team can comprise the owner or the contractor separately or both in combination. It is also possible to include other stakeholders such as subcontractors and suppliers.

Risk scope includes administrative, commercial, construction, logistical, contractual, engineering, environmental, health and safety, project management, quality, and security risks.

There are different risk levels. Strategic, tactical, and operative risks are internal risks of the megaproject. Industry and the PESTLE framework describe external risks (cf. Section 9.1.3). PESTLE stands for political, economic, social, technological, legal, and environmental risks (Figure 8.11). Internal risks can influence each other.

Risk governance describes the overall approach to risk management. Does the owner prefer an exclusive or inclusive approach? The exclusive one concentrates risk management within one stakeholder group, e.g. the owner's risk management team. The inclusive one creates joint risk management teams where two or more stakeholders join their efforts; this could be a team consisting of members representing the owner and the contractor. In this way, it is possible to internalize interface risks.

The risk management process starts with a review of the objectives and moves on to risk identification, analysis, evaluation, responses, reporting, reviewing, and monitoring. Part of

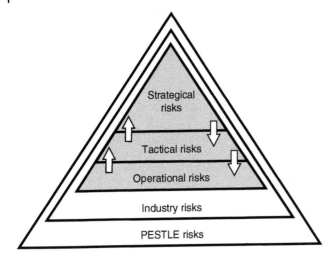

Figure 8.11 Risk levels.

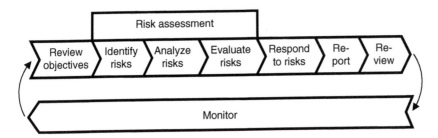

Figure 8.12 Risk management process.

reviewing and monitoring is a continuous improvement process that includes a review of objectives, e.g. the risk appetite. Risk analysis and evaluation can rely on qualitative or quantitative methods (Figure 8.12).

Risk responses include avoiding, transferring, mitigating, or accepting for negative risks and exploiting, sharing, enhancing, or accepting for positive risks.

Risk communication provides chances to create awareness and train employees and workers to identify and report risks.

8.2.9 Procurement Management

The importance of procurement depends on make-or-buy decisions and on the relative importance of materials. If the decision of the general contractor is to subcontract the largest part or all of the works, then procurement becomes highly important. The same holds true for developing economies and those in transition, where materials make up around 70%–80% of the contract sum. As previously described, this is due to the low labor wages of around 1 USD per hour.

The make-or-buy decisions determine the supply chain, and accordingly, supply chain management becomes important. Typically, procurement of materials and subcontractors relies on global sourcing. The megaproject contractor must have access to a global network of product and service providers.

Lean construction can be very helpful for organizing seamless and just-in-time logistics.

8.2.10 Stakeholder Management

For each project, we need to identify the stakeholders and manage their expectations. Stakeholders come from different areas and typically form groups. Forgetting a group of stakeholders might prove disastrous.

 Stakeholder identification = (obs.) Stuttgart 21 is a megaproject in Germany. The owner is Deutsche Bahn (German Rail), and they are building a new central station in Stuttgart with tunnels transforming it from a terminus to a through station. The planning stage followed legal requirements, and the public had a chance to check the plans and raise objections. Deutsche Bahn worked off all objections, and the project began. At the time, large public groups began demonstrations against the project and pressured politicians. Only a positive vote in the state parliament made it possible to continue the project.

Deutsche Bahn did not identify a disgruntled public as a stakeholder, and the project was very close to termination.

We can organize stakeholders in several levels. Figure 8.13 shows the top level for an infrastructure megaproject in Bangkok, the elevated BangNa expressway. The Thai government oversaw three ministries involved in the project. The Expressway and Rapid Transit Authority (ETA) acted as a direct owner, signing a contract with the ICJV. ETA reported to the Ministry of Interior and was the only authority with a right to build and operate toll roads. The Department of Highways (DoH) did not have this right and could not act as the owner. As ETA planned the elevated expressway above the median of an existing highway under the control of DoH, this authority was in possession of the right-of-way. DoH reported to the Ministry of Transportation. The Ministry of Finance was involved in financing the expressway. It issued promissory notes to the contractor who had signed a design/build/finance contract with the owner (ETA) and an additional financial contract with four Thai banks. The contractor handed the promissory notes over to the banks and received cash for construction. At the end of a construction phase, the contractor handed the completed structure over to the owner, and the Ministry of Finance redeemed the promissory notes. Diverse insurance companies provided coverage for different aspects of the project.

Utility relocation is a major task for many projects. In this case, electrical, water, telephone, and gas lines needed relocating. The contractor had to get approval from the concerned authorities.

Two German companies (Bilfinger Berger and Dywidag) and one Thai company (Ch. Karnchang) formed an ICJV and signed the contract with the owner. They also signed

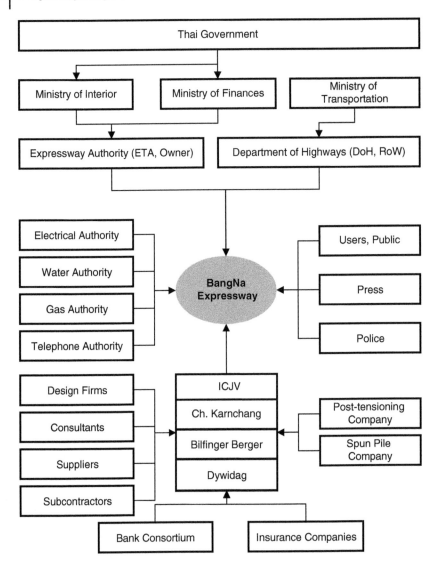

Figure 8.13 Identified stakeholders on the first level for a megaproject in Bangkok.

numerous contracts with design firms, consultants, suppliers, and subcontractors. The three partner companies founded two independent companies whose initial contract was with the ICJV for post-tensioning supplies and prefabricated spun piles. These companies initially relied completely on the very substantial contracts with the ICJV.

The expressway was built over continuous traffic; accordingly, the police had high road safety concerns. Construction affected the users of the at-grade highway, and the press followed the project closely. All stakeholders joined to form a multi-organizational enterprise (MOE). It might be clear that the expectations of the different stakeholders are divergent.

It will never be possible to fulfill all expectations. Therefore, it is necessary to distinguish very important from not so important ones.

8.2.11 Health, Safety, and Environmental Management

HSE management does not have the same importance around the globe. This is true separately for health and safety as well as the environment. Therefore, local regulations and norms are of paramount importance.

In many countries, there remains a gap between external communication and internal practices: internal practice does not always live up to external talk. After lessons learned over years, I would always place the safety of workers above profit, and there are unfortunate payoffs between the two. Regardless of norms in one country, we can always resort to individual action.

8.2.12 Contract Management

The construction contract is the most important document for a project, be it local or international, small or large. In addition, contracts for megaproject are highly complex. Oftentimes, owners use standard contracts. The World Bank has, for example, agreed to use FIDIC contracts (Fédération Internationale des Ingénieurs Conseil, the international federation of consulting engineers). There are six types of contracts: (i) design/bid/build – "Red book," (ii) design/build – "Yellow book," (iii) EPC Turnkey – "Silver book," (iv) consultant contract – "White book," (v) design/build/operate – "Gold book," and (vi) short form – "Green book." FIDIC contracts have a section with general conditions that owners should not change and another one with particular conditions that allow adjusting to the circumstances of a specific megaproject.

The understanding of the purpose of construction contracts differs between the East and the West. The Western understanding builds on legal traditions reaching back to Roman times. The Latin phrase *pacta sunt servanda* (contracts must be served) describes the approach. The focus is on the offer and acceptance of a contract. The Eastern understanding stresses adaptability, i.e. contracts should be adapted to changing circumstances. The focus is on the meeting of minds and the intention of the contract. This way, a confrontation between a rigid understanding and a flexible one might lead to conflicts. Contractual flexibility coupled with the pursuit of self-interest can lead to confrontation as well as a rigid insistence on contractual arrangements.

All complex contracts are incomplete. The theory of incomplete contracts discusses several reasons for this. Unforeseen contingencies, cost of writing, and costs for enforcing complete contracts are easy to understand. It is almost impossible to foresee all contingencies for complex megaprojects. This follows from the prior statement that it is impossible to define even the project scope completely. In addition, there exist numerous unknown conditions (e.g. precise knowledge of soil conditions) or uncontrollable influences, such as inclement weather. If it would be possible to foresee all contingencies, it would still be too time-consuming and too costly. Enforcement of the contract by the owner is as difficult as

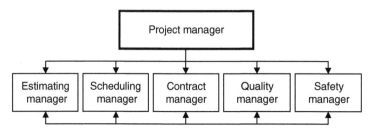

Figure 8.14 Organization chart with contract manager.

by courts in case of escalation (Tirole 1999). The economic discussion of the reasons for incomplete contracts is much more detailed, but for our purposes, this shall suffice. Since construction contracts are incomplete, we need a tool for adjusting them to the prevailing circumstances. This tool is the unilateral right of the owner to demand changes while bearing the consequences with regard to time, cost, and quality. This is accompanied by the right of the owner or the contractor to claim compensation for failings by the contractual partner.

Several contracts require attention in an ICJV. These are (i) construction contract, (ii) joint venture agreement, (iii) design contracts in case of design/build, (iv) consultancy contracts, (v) purchase contracts, and (vi) subcontracts. Purchase and subcontracts especially tend to be numerous.

Few must be the ICJVs that do not employ a dedicated contract manager – I have never heard of even one. The main responsibility for the contract rests with the project manager, and the contract manager reports directly to him. The contract manager will have to stay in close contact with the estimating, scheduling, quality, and safety manager (Figure 8.14). Often, contract managers are trained as quantity surveyors. Many contract managers work as freelancers, and an ICJV might employ them as third-country nationals.

Typically, the owner formulates the construction contract. The tasks of the contract manager include the understanding and interpretation of this contract, reaction to changes, formulating legal letters, and writing invoices. The partner companies to the ICJV specify the joint venture agreement, and again, the contract manager must understand and interpret it. The contract manager must write the other contracts (design, consultancy, purchase, and subcontracts). He will then have to deal with changes and invoices. Contracts are so important that the project manager must always stay involved and make the final decisions. For all these activities, the contract manager needs information on cost, time, quality, and safety, and he must keep the responsible managers informed.

In some Asian countries, owners refuse to grant additional payments for justified claims even when agreeing to a time extension. This is not logical. The contractor determines the resources required for a cost minimum for a given construction time. He then provides exactly these resources to the megaproject. Changes to the contractual construction time lead to cost changes because of over- or underutilization of the resources provided. It is possible to add or reduce the resources, but this costs additional money. Cost and time curves have a U-shape when the resources are constant (Figure 8.15).

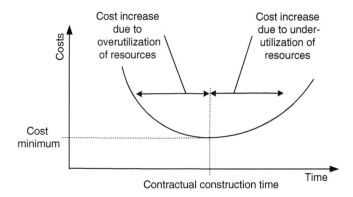

Figure 8.15 Cost and time relationship for constant resources.

8.3 Production Planning

Production planning includes all activities that provide the necessary information for construction. The bidding team produces basic information before signing the contract at a level sufficient for submitting a bid. This level is not adequate for construction, and more detailing and final decisions are required.

The most abstract way of thinking about production is in the form of a production function, which shows the dependency of output on input. In the simplest case, two input variables, x_1 and x_2, determine the output y:

$$y = f(x_1, x_2).$$

In construction, these inputs could be labor and equipment with the materials staying constant. A substitutional production function can approximately describe typical construction projects, i.e. equipment can replace labor and vice versa. Figure 8.16 illustrates such a production function for three output levels y_1, y_2, and y_3. Input x_1 represents labor, and input x_2 represents equipment. While the input ratio (the slope of the output line) is the same for the output levels y_1 and y_2, it is different for y_3. It is important to note that the

Figure 8.16 Substitutional production function.

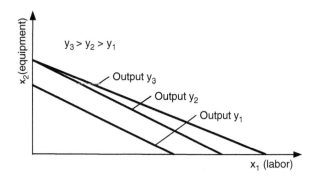

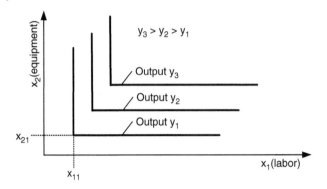

Figure 8.17 Limitational production function.

ratio between labor and equipment is not fixed. Comparing the output y_2 and y_3, we can find that the higher output y_3 is due to an increase in labor only (x_1).

This is different for limitational production functions (Figure 8.17). Here, the input factors for one output level are fixed. Adding more of one input without adding proportionally more of the other input will not increase output. A simple example is a table with four legs. With eight legs and two tabletops, we can produce two tables. Adding 10 legs does not allow us to produce more tables.

The combination x_{11} and x_{21} is the only efficient one for output level y_1. Jumping to the next output level y_2 means using $n * x_{11}$ and $n * x_{21}$, if $y_2 = n * y_1$, with n being a fixed number. Limitational production functions especially characterize infrastructure megaprojects.

Limitational production functions describe not only megaprojects but also production planning for lean construction. The crews for the different trades in a high-rise building are fixed in size so that they finish work in a certain area (module) at the same time. This allows for establishing cycle times.

 Limitational production function = (obs.) The construction technology for the BangNa Expressway was segmental construction for the superstructure. In this case, the contractor precast bridge segments in a yard using molds. After production, the concrete segments gained their 28-day-strength in storage areas in the yard. The contractor shipped them to the erection site by using trailers with just-in-time delivery. Then, he placed them on a girder by using a crane. Finally, he pre-stressed them to form a uniform bridge span (Figure 8.26). The methodology required a fixed ratio of molds, storage areas, trailers, and erection girders. The time schedule determined the use of five girders with a maximum monthly capacity to place 1105 segments. Accordingly, 22 trailers allowed transporting 1030 segments and 48 molds to produce 1118 segments. Only the storage areas had a higher capacity (1650 segments) to serve as a buffer between production of the segments in the yard and their erection on site. Figure 8.18 also shows that the erection girders follow the pull principle of lean construction with the erection girders exerting the pull.

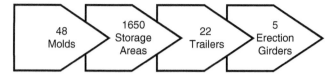

Figure 8.18 Limitational production function for the BangNa Expressway.

> The chosen construction technology allowed erecting 2500 m of superstructure; this is extremely fast. However, increasing the speed would require a lead time of 6 months for the production of the resources. One additional girder, 5 trailers, 330 storage areas, and 10 molds would have increased construction speed by 20%. This production function had a very high output but little flexibility. Substitutional production functions, on the other hand, have high flexibility but limited output.

Besides the decision for a production function, the contractor must choose a production approach. Choices include own work versus subcontracting, on-site versus off-site production, and traditional versus lean construction. Own work presupposes the in-house technological know-how and offers better control; off-site production can almost double the construction speed and increase quality while requiring high investments, and lean construction can increase speed, lower costs, and improve quality. The basics for lean construction during production planning are involvement of all participants in the supply chain, modularization and standardization, as well as precisely timed repetitive processes (cycles). Taking the BangNa Expressway as example for the application of lean construction, one bridge span served as module. Precasting provided standardization and the cycle time for one bridge span as a module was 2 days. Accordingly, the contractor harmonized all resources to complete segment production, storage, transportation and erection in 2 days. This lead to the fixed ration of resources establishing a limitational production function. Lean construction focuses additionally on customer satisfaction, reduction of waste to increase productivity, and continuous improvement.

Even the traditional process of production planning is highly complex and requires much input. For each solution, we must harmonize 11 partial plans as they are interlinked. The planning process is iterative and non-sequential, i.e. it is a circular process and a circle has no beginning. It is up to the planners to determine a start.

The partial plans include:

1. make-or-buy decisions (own work or subcontracting)
2. scheduling (determining how to finish on time)
3. estimating (establishing a budget with details on items as a benchmark for cost control)
4. logistics (transport, storage, and workflows)
5. legal organization (contractual processes)
6. HSE organization (safety and environmental standards)
7. management organization (organization chart and rules)
8. site installation (factory layout)
9. construction technology

10. resource planning (manpower, material, and equipment)
11. quality planning

All 11 plans depend on the design either directly or indirectly. This means that engineers must follow the impacts of every design change throughout all plans. However, not only design changes but also process changes affect the partial plans. If engineers innovate to develop a better construction technology than previously assumed, it will certainly require changes to the schedule, budget, logistics, site installation, resource, HSE, and quality planning. Likely, the management and legal organization also need adjustment (Figure 8.19).

Production planning starts at one point, e.g. construction technology, and immediately needs input from other plans such as scheduling and resource planning. Therefore, the engineers switch from one plan to another to get the information required at a stage; this is a non-sequential process. Often, late information makes changes to previous decisions necessary; this is an iterative process. If we now consider scope complexity, it becomes clear that the process of production planning for megaprojects cannot be free of mistakes. In such cases, it becomes more important to be comprehensive than to be exact. Being exact

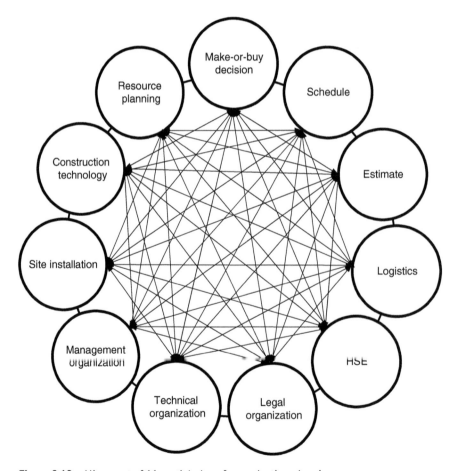

Figure 8.19 Alignment of 11 partial plans for production planning.

Figure 8.20 Profitability of high investments.

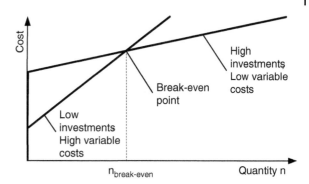

brings the danger of getting lost in details. Being comprehensive directs attention toward the broader scope.

Production planning is, in many cases, the area where contractors display their innovative potential. In an ideal world, the contractor with the best production planning should be able to offer the owner the lowest bid, and the owner should accordingly award the contract to the smartest contractor. In the real world, the winner is often the contractor with the largest estimating error (winner's curse).

Megaprojects allow for large investments, i.e. specific solutions. The higher the initial investment as fixed costs, the lower the variable cost should be. If this does not hold true, the solution is a very inadequate one. If it holds true, then it depends on the quantity produced whether the high investments for specific resources are advantageous. The break-even point between low investments/high variable costs and high investments/low variable costs depends on quantity. The higher the quantity, the more advantageous are higher investments. The literature gives a few examples of innovations in megaprojects. Brockmann et al. (2016) identify 58 innovations for the BangNa Expressway; some of these are product or construction technology innovations, and others are within the technical, management, or contractual organization. This is certainly an impressive amount, and there is no indication why the BangNa Expressway should be outstanding. Abbott et al. (2007) speak about hidden innovation, and it is true that innovation is hidden to researchers as well as project participants. The latter group lacks an operational definition of innovation, and is therefore unable to identify innovation except on the highest level; the former does not have sufficient access to information. My assumption is that many megaprojects are as innovative as the BangNa Expressway. This is possible because of the large quantities involved, and it becomes necessary because of the singularity of megaprojects and competition in global markets.

Megaprojects always allow for higher investments in specific resources. Accordingly, innovation based on specific resources in new processes becomes a viable option (Figure 8.20).

8.4 Site Installation

Site installation is a topic that has drawn little attention. The problem lies in the fact that most sites demand different approaches and the available areas are very different as well.

Accordingly, there are no general rules for developing site layouts except on an abstract level (place a tower crane in a way that minimizes boom, etc.). For the discussion of site installation, we need to refer to examples. I have chosen again the BangNa Expressway.

The site installation is the factory where we produce our structures, similar to how carmakers produce cars in a factory. The comparison also highlights the differences between the two sectors. A specialized team minutely develops the production line for car production. The basic units for each step of the process are minutes, and each step receives exactly the necessary amount of labor and equipment. This is possible because of large output from the factory, which also allows for very specific investments such as robots. Even for a megaproject, the output from one site is relatively small as the construction industry has to set up their factories for each project separately (the industry of mobile factories). Megaprojects allow for more investments in site installation but never as much as in the automobile industry.

The shape of the area available for site installation is never ideal; in inner cities, it is close to catastrophic from a production view. There are projects which only offer space for site installation in the growing structure; the factory is set up in the product. Infrastructure megaprojects often allow for larger available areas. However, they are also not ideal. Figure 8.21 shows the layout of the precast yard of the BangNa Expressway. It resembles a car factory much more than most other site layouts and is still deficient.

The ideal layout for a precast yard would be in the immediate vicinity of the erection site. Given the length of the BangNa Expressway – 55 km – this would be at 27.5 km. The shape would be quadratic. Switching from ideals to real life, we need to buy or rent land at a low price.

The precast yard for the BangNa Expressway in BangPo comprised 650 000 m^2; this is a lot of land. It belonged to different people, and some might have no incentive to sell. In the process of buying, the price will rise as some landowners will speculate that their land becomes indispensable. In the end, the land for site installation – be it for off-site or on-site construction – will most likely have an irregular shape (Figure 8.21). The precast yard lay at a distance of 4 km from the expressway, again not ideal. It meant to strengthen the road leading to the highway with a 20 cm concrete top layer for the heavy traffic despite the fact that loads did not exceed Thai traffic regulations.

Besides the precast yard, other site installations included three site offices with space for work and storage plus a main office. The contractor was able to find space for these alongside the 55 km of construction area.

There are two main features of the precasting – the human and the technical environment. The following numbers in brackets refer to Figure 8.21. Two thousand workers lived in the labor camp (11), some with their families. This is the size of the village, and all the necessary services need to be provided. In countries like Thailand, there will always be enough initiative to set up a market in the vicinity and a lack of regulations to stop it. Vendors established such a market across the road; it existed as long as there were workers in the precast yard. The workers could also use the canteen in the yard (13).

The design specified three types of segments. The largest amount (22 000) were wide-spanning D6 segments with a width of 27.2 m. D2/D3 segments with a width of 15.6 m and portal segments were less numerous. Producing 22 000 segments almost allows for the installation of a production line. Figure 8.22 displays a schematic workflow.

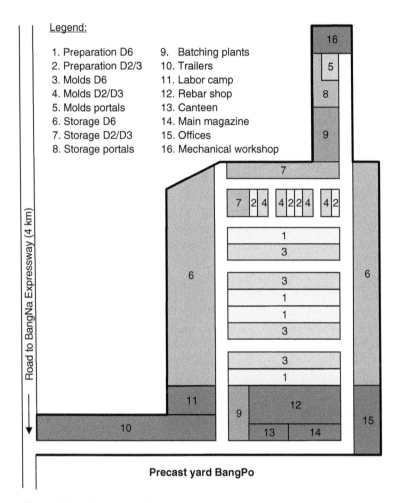

Figure 8.21 Precast yard for the BangNa Expressway in BangPo.

In Figure 8.21, there are three areas dedicated to D6 segments: preparation (1), production (3), and storage (6). Two batching plants (9) mixed the concrete, a subcontractor cut and bent the rebars for the cages (12), and some further parts (transversal post-tensioning, deviator forms, and struts) were prepared in a small area close to the mechanical workshop.

The design determines the workflow (and vice versa). The D6 segments are 27.20 m wide, 2.60 m high, and 2.55 m long. Internally, two struts with a cross-section of 25 by 25 cm strengthen the segment. Four transversal tendons run through the top slab and two from the top slab through the webs and the bottom slab back to the top deck. On-site, construction crews placed and stressed the longitudinal post-tensioning. For this, the segments had deviator blocks through which the on-site crews needled the tendons. The molds (12 molds in 4 rows for a total of 48) gave the segments the designed shape (3).

Scheduling determined the number of molds based on the assumption that it would take a day to produce one segment. An exception was the pier segments, which are more solid to allow space for the anchoring of the longitudinal tendons; their planned production cycle was 2 days. This assumption made it impossible to assemble the rebar cage in a mold, as it took 13.5 hours to assemble the cage and 18 hours to pour the concrete, including 10 hours for hardening. The sum exceeds the 24 hours in a day. Therefore, a preparation area for the production of the rebar cages became indispensable (1).

In a first step, the struts and transversal tendons are prepared in their respective areas and transported to the preparation area on trucks. The rebar crews cleaned deviator forms next to the mock-ups for the rebar cages. Next, they tied the rebars together, including struts, deviators, and transversal tendons. A tower crane lifted the complete cage into the mold (Figures 8.22 and 8.23).

The production of the segments followed the match-casting technologies. The segment cast the day before served as part of the formwork for the next segment. They were cast against each other to assure perfect fit during assembly on site. Accordingly, the first step was the preparation of the mold after shifting a segment out of it. The molds required cleaning and surveying, as most segments are slightly different because of the alignment of the bridge.

Once the molds were ready, the rebar cages were lifted in, and truck mixers delivered the concrete from one of the two batching plants. The truck mixers dumped the concrete on conveyor belts, which transported the concrete into the molds. After 10 hours of hardening in the form, crews stressed two of the transversal tendons for partial post-tensioning before shifting the segment out of the mold. The bottom formwork of a mold sat on rails, and the segment could be pulled out of the mold into the match-cast position. A day later, it was moved to the curing position. Each mold needed three bottom formworks, one for

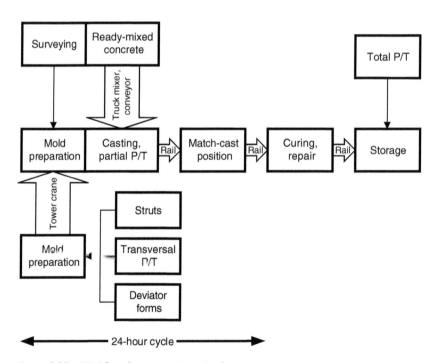

Figure 8.22 Workflow for segment production.

Figure 8.23 Complete rebar cage with D6 molds. Source: Christian Brockmann.

casting, one for the match-cast position, and one for the curing position (Figure 8.24). From here, shuttle lifts transported the segments into storage for reaching 28-day-strength. After reaching this concrete strength, crews applied the full post-tensioning of the segments in the storage areas. Figure 8.25 provides an aerial view of the precast yard.

To summarize the important points of this subsection, I want to direct attention to the following points:

- Planning of site installation depends on the design and the other partial plans.
- The ratio between the cost for site installation as input and the quantity produced even in megaprojects as an output is small in comparison with the factories of other industries, e.g. the car industry. This prohibits planning efforts and investments in the same magnitude as in the car industry.
- The larger the project, the higher the return on planning input.
- The larger the project, the more cost effective is the investment in site installation if it is possible to lower the variable costs. The investment cost for the BangNa Expressway

Figure 8.24 Production line for D6 segments. Source: Christian Brockmann.

Figure 8.25 Aerial view of the precast yard in BangPo. Source: Christian Brockmann.

totaled 150 million USD – certainly an impressive amount. Given the large production quantity, investments in site installation will lower the overall costs.

- Planning of site installation is a very demanding task and requires special production knowledge in the case of megaprojects.
- Contractors use methodologies similar to the car industry if the amounts produced are large.

8.5   Construction

Construction requires the largest amount of resources from all activity groups. This holds true for manpower, materials, equipment, and subcontractors. Accordingly, managing and physical work dominate this phase.

 Construction = (def.) All activities to transform construction planning into the contracted structure. It includes management with the activities of planning, organizing, staffing, directing and controlling. The focus lies on directing and controlling as well as on short-term planning and organizing with a horizon of 1–2 weeks. Long-term planning is a part of production planning. The most important aspect of staffing is the daily motivation of engineers and workers.

Megaproject construction has to deal with very large quantities and very often with cutting-edge technologies. These two elements combine to create high complexity, especially as construction time is very often short.

Taking the BangNa Expressway again as example, the contractor had to place 1.8 million m^3 of concrete, 180 000 tons of steel, and 50 000 tons of post-tensioning. The highest building in the world – itself certainly a megaproject – needed 330 000 m^2 of concrete and 33 000 tons of steel, i.e. the quantities of the BangNa would have been sufficient for six buildings the size of Burj Khalifa. It took 6 years to finish Burj Khalifa; 3 years had to suffice for the BangNa Expressway. However, it is certainly easier to build horizontally than vertically. The question is whether this is still true when tens of thousands of cars pass through the construction area on a daily basis. Each megaproject has to deal with its own difficulties.

In segmental construction, we can use overhead or underslung girders to place the segments. The use of overhead girders is widespread, and they still continue to pose new challenges for each project. Underslung girders are a more innovative approach. Figure 8.26 shows the underslung girder from the BangNa Expressway. A swivel crane mounted to one end of the underslung girder lifts a D6 segment produced in the precast yard in BangPo and transported by specially designed and built trailers to the erection site. It is visible that the whole production is highly specific and innovative. At the beginning of the project, nobody had any experience designing and producing the new D6 segments. Nobody knew the best production process in the yard, and nobody had experience transporting the segments.

Figure 8.26 Segmental construction with an underslung girder. Source: Christian Brockmann.

Nobody on the job had ever worked with an underslung girder nor had any experience with the erection process.

Asset specificity and the willingness and ability to learn together with a high level of related experience are the key to success. As there are thousands of engineers and workers involved in construction, spreading and anchoring the knowledge is as important.

 Implementing innovative technology = (obs.) The Expressway and Rapid Transit Authority as its owner contracted an ICJV for the construction of an elevated expressway (not the BangNa expressway). This ICJV used so much time to master the technology of segmental construction with an overhead girder that progress was very slow. The owner lost confidence and terminated the contract. They then signed a contract with another ICJV, which also had the same problems at the beginning. Again, the owner was close to losing patience, when finally progress accelerated. In the end, it took 2 days to erect a bridge span. The same ICJV took on the BangNa expressway with a different technology by using segments twice as wide and underslung girders. Reaching a 2-day cycle did not take long; the ICJV had the necessary operational expertise.

The previous observation shows that learning is of paramount importance to implement cutting-edge technology. This learning takes place in steps by introducing planned changes (induced learning) or incrementally by day-to-day learning (autonomous learning).

On the basis of a sound design and a well-functioning overall organization, the contractor will be successful when mastering the technology and the logistical challenges.

 Construction and complexity = (obs.): The ICJV of a tunnel project comprised two French, one American, one Danish, and one German contractor. Based on their experience from the Channel Tunnel, the French companies took responsibilities for the tunneling with tunnel boring machines (TBMs). After months of no progress, the JV board decided to entrust the American company with the same responsibility. After many more months and no progress, they next named the German company to head the tunneling efforts. The Germans succeeded.

Does the story tell us that German engineers were better at tunneling than their partners were or that the decisions made by the French and American managers and engineers finally bore fruits?

For me, this is another example of a slow complexity reduction at the beginning of a project and a nervous JV board. I think the French engineers would also have been successful at the time when the German engineers received praise.

9

Management Functions

Many business schools promote the idea that managing consists of a set of ideas, tools, and approaches that promise success under all kinds of circumstances. I do not agree. Construction project management demands its own approach and so does each megaproject, i.e. I advance the idea of a contingent approach. It depends on the level that we are looking at. Overall, a generic management approach will be successful; in detail, we need to apply a specific approach considering the special context.

Figure 9.1 reminds us of the place of management functions in the descriptive megaproject management model.

The discussion of the overall management process will follow five activities or management functions: planning, organizing, staffing, directing, and controlling.

Certain characteristics of the activity emerged when discussing management in general in Section 3.2. The most important ones that also apply to megaprojects are:

- Managers communicate more than 70% of the time, often listening.
- The idea that managers first make a plan, then create an organization, staff it, tell the people what to do, and finally control results is wrong. Management is not a clean sequential process.
- Management is messy, and managers must often take decisions without having all the desirable information.

Many young managers feel uncomfortable with the transition from engineering to managing. They complain that they did not accomplish anything during the day, that others just kept them engaged in communication. However, managers in megaprojects receive their paychecks exactly for talking and listening. Engineering concentrates on tasks; managing focuses on people. Of course, there is a management hierarchy in a megaproject: the closer a manager is to the operative level, the more important the tasks become. The project manager will almost entirely use communication to deal with and generate information, to lead and link people, and to deal with others.

Most people, at least in the Western world, prefer sequential work. Megaprojects are the worst place to look for such processes. Typically, project managers, even of small construction projects, profess that they find time for planning only after everyone else has left the construction site. In megaprojects, there are always some people at work, and the project manager might find no time for planning. Engineers and lower-level managers must

Advanced Construction Project Management: The Complexity of Megaprojects,
First Edition. Christian Brockmann.
© 2021 John Wiley & Sons Ltd. Published 2021 by John Wiley & Sons Ltd.

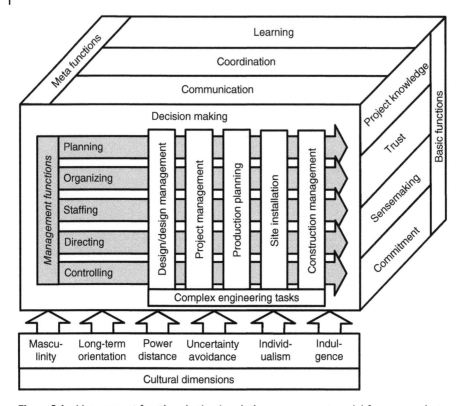

Figure 9.1 Management functions in the descriptive management model for megaprojects.

prepare the plans while top management approves them. Others often dictate the order of work, and it is a constant concern of top management to find time to be proactive instead of reactive. The reactive pressure is extremely high and leads to highly fragmented work processes.

 Planning = (obs.): Especially at the beginning of the BangNa project, I found no time for planning and developing ideas. The pressure woke me up most nights and I made a point to wake up entirely. Then, I was quite happy to stay comfortably in bed, undisturbed. This was the best time for planning. When discussing this with other project managers, they told me of similar behavior.

The workload at the beginning of a project is overwhelming and it stays high throughout. Managers can only reduce the workload by making decisions. Yet, there is little information at the beginning of the project. Decision-making without the desirable information requires shifts from rationality to intuition based on experience. Some of the decisions will quite naturally be wrong, and the top management cannot afford to be afraid, since the situation is messy. It becomes messier when superiors come to judge the decisions half a year later and take into account the information available at this time. Prima facie, the top management looks bad; actually, the superiors demonstrate a lack of task and social understanding.

 Decision-making = (quote) "…most decisions should probably be made with somewhere around 70% of the information you wish you had. If you wait for 90%, in most cases, you're probably being slow. Plus, either way, you need to be good at quickly recognizing and correcting bad decisions. If you're good at course correcting, being wrong may be less costly than you think, whereas being slow is going to be expensive for sure."

Bezos (2016)

Planning and controlling as well as organizing and directing are twin functions; they condition each other. I will present the functions in a sequence different from the descriptive megaproject management model so that I can discuss the twin functions next to each other. The order then becomes as follows: planning, controlling, organizing, directing, and staffing.

9.1 Planning

Plans should have the following characteristics:

- Completeness (accounting for all important facts)
- Relevance (distinction between important and unimportant)
- Accuracy (relative exactitude)
- Topicality (newest available data)
- Objectivity (judgment is delayed until decision-making)
- Flexibility (adjusting to a dynamic environment)
- Clarity (unambiguous)
- Feasibility (no dreams)
- Consistency (no contradiction between plans)
- Goal focus (directed toward preset goals)
- Efficiency (no waste of resources)

It is easy to understand that no plan will ever meet these characteristics. However, they provide planners with a direction. A planner must often use his expert judgment to decide between conflicting demands.

Planning and controlling are twin functions because the plan is the cognitive ideal that we later compare with the actual situation. On the one hand, without a plan, there is no benchmark. On the other hand, we cannot be sure to follow the plan without control. This sounds self-evident. Unfortunately, there are many examples when managers do not follow the logic. The reason is a lack of responsibility and an abundance of self-protection. Plans are, foremost, forecasts of the future. As nobody knows the future, plans must more or less be incorrect. When they are very wrong, powerful managers might stop controlling in order to hide their own planning mistakes. At other times, the implementation of plans might be so defective that, again, powerful managers might neglect controlling or even cheat. This is in no way admissible; it will endanger the success of the project by putting self-interest first.

There are many books on strategic planning; a good one is by Mintzberg (1994), titled *The Rise and Fall of Strategic Planning*. Other authors think much better of strategic planning. In construction, Langford and Male (2001) discuss strategic planning with regard to the construction company and not to projects. To restrict strategic planning to the company level is not convincing because the annual turnover in megaprojects often exceeds that of construction companies, except for the largest ones. It helps to demystify strategic planning by classifying it by its planning horizon. We might distinguish between two planning horizons that are relevant for megaprojects – strategic long-term and operational short-term planning. There is no need for a third intermediate level, often discussed as tactical planning. Strategic planning sets a frame for operative planning without predetermining it. There is freedom for operative planning. Table 9.1 shows some important differences between these horizons.

A schedule can be a strategic or an operative plan. Project management will work out an overall schedule for the project, and construction management (on the level of area managers or supervisors) must detail this overall schedule as a weekly or bi-weekly schedule. Area managers and supervisors are very close to construction activities on a daily basis and have much more detailed information for a short-term schedule. For this reason, it is a not a good idea to develop the weekly or bi-weekly schedule in the project management office. The uncertainty is undoubtedly higher for the overall schedule. When creating a schedule, there is absolutely no structure. In many cases, only the starting and ending points are known; everything in between is terra incognita. Charting this terra incognita is a case for scheduling specialists. A supervisor as construction specialist must respect the overall schedule into which he will embed his bi-weekly schedule. Construction duration for a megaproject is long; 2 weeks are short. The terra incognita between the start and end provides for many alternatives, and this is not true to the same degree for a bi-weekly schedule. A bi-weekly schedule must contain all activities in detail; this is the main purpose for

Table 9.1 Characteristics of strategic and operative planning in megaprojects.

	Strategic planning	Operative planning
Hierarchical level	Top management	All levels
Uncertainty	Relatively large	Relatively small
Kind of problems	Unstructured and complex	Relatively well-structured and repetitive
Time horizon	Long term	Short term
Sources of information	From the environment	From within the project
Alternatives	Many alternatives	Some alternatives
Scope	Important aspects	All aspects
Detailing	Global statements	Detailed statements

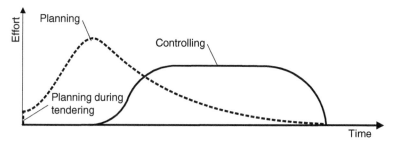

Figure 9.2 Planning and controlling over the project duration.

creating one. Neither is a strategic plan grandiose nor is an operative plan simple; both serve their separate purposes.

As we are defining strategic and operational goals for megaprojects through planning, it is clear that most planning activities take place in the first part of the project. At least operational planning is continuous to the end. Sometimes, it also becomes necessary to revise strategic goals. Controlling can only start once plans are available, i.e. after some plans allow for comparing planned versus actual. Figure 9.2 shows the typical curves for planning and controlling without an adjustment of the goals.

9.1.1 Analysis

Plans are often in the form of a SWOT analysis, where the planner looks first at the international construction joint ventures (ICJV) with its strengths and weaknesses (SW), then at the environment with its opportunities and threats (OT). Taking the example of a schedule, a SWOT analysis can serve to prepare the actual schedule by naming the assumptions that form the basis of the schedule. It is better to write these down than to keep them in the mind of the planner. The SWOT analysis discusses, on a meta-level, the context of the schedule. It is also a first step to schedule risk management. These statements hold true for most plans, not just schedules.

A comparison of internal SWs with external OTs allows generating different options. There will be pros and cons for each option, and an evaluation of these will lead to a choice of guiding assumptions to be used as a basis for a particular plan. Implementing the choice means to consider the assumptions when developing a plan such as a schedule (Figure 9.3).

The implementation of a new construction technology is common for megaprojects and it poses threats to a schedule. SWs may be the skill of the workforce, relevant experience of managers and engineers, the ability and motivation to learn quickly, and access to external advice. The physical environment with limitations of space and accessibility, project duration, availability of materials, and the attitude of the owner as well as other external stakeholders may pose OTs. There are many more possible influences; it depends on the specific situation.

9.1.2 Developing New Plans

The singularity of megaprojects requires fresh approaches, among them new solutions in planning. Someone might imagine that a stroke of genius typically provides us with new

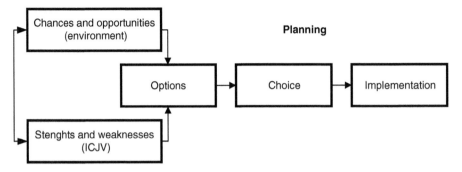

Figure 9.3 SWOT analysis for planning.

ideas; this is dead wrong. In most cases, hard work is the foundation. As Edison is quoted to have once said, "*Genius is one percent inspiration, ninety-nine percent perspiration.*"

Sometimes, a stroke of genius actually is the source of an original approach when developing a plan. In most cases, the planner will pick up older plans and compare their approach, strength, and weaknesses. These plans can be company plans or available through open resources such as the Internet. It is worthwhile to place a number of plans from different sources next to each other and look for differences. A lazy planner will copy one available plan with minor adjustments; the result will not be good. A diligent planner will compare all available plans and elaborate a SWOT analysis. He will then insert variations into the plan by asking the "what if" question. The more persistent the planner is in his effort, the more novel and applicable the plan can become; this is a process of developing cognitive complexity. The "what if" question is a very powerful tool if coupled with imagination. The next step is to check the feasibility of a solution; here, the planner has to determine whether it is part of the ICJV's technology space. If it is, he might break off the search for a solution because the one found exceeds his own expectations. This is the application of a satisficing heuristic. Before a final decision on the plan is made, there will be, in most cases, an internal discussion with the superiors and others involved. The solution must also surpass the expectations of this group of decision-makers (Figure 9.4).

I think it is a good idea to spend half the available time to develop a structure for a plan, and the other half filling the structure with content. The "what if" question drives the structure more than the content. It is also not helpful when too many people are involved in decision-making for a specific plan. The more people, the more demands that lead to inconsistencies. However, the relevant people must be part of the planning process; foremost among them is the project manager. He must take responsibility for setting strategic and some operational goals.

9.1.3 Analytical Framework for Planning

A systematic approach to a SWOT analysis is possible with the help of frameworks providing categories. One such a tool is the PESTLE framework for a context analysis. PESTLE is an acronym for the political, economic, social, technological, legal, and environmental context of a megaproject. This framework might be too ambitious for an office building, but not for megaprojects. In the end, some megaprojects can influence the economy of a country,

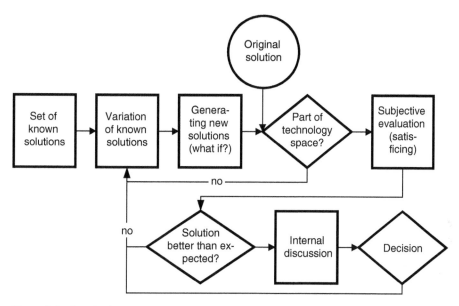

Figure 9.4 Developing a new solution for a plan.

such as the Aswan Dam on the Nile in Egypt. Megaprojects can also bankrupt construction companies.

The project environment affects a megaproject alongside the categories of the PESTLE framework. Most important is the owner, but consultants, suppliers, and subcontractors also play a role, as well as institutions. There is a very stark institutional difference for megaprojects in democratic and semi-democratic or authoritarian governments. In democratic societies, preparations for megaprojects take decades; in authoritarian societies, this takes a year or even less. The difference remains important during implementation.

Figure 9.5 demonstrates the approach to strategic planning for megaprojects. The environmental analysis focusses on the PESTLE and project environment; the internal analysis pays attention to the resources of the ICJV. In a second step, options arise from the analysis through further work as described above. The choice follows satisficing behavior and a critical evaluation.

9.1.4 Planning System for Megaprojects

Quite a large number of plans might be required for a megaproject, and these must form a consistent system. Ideally, there should be no goals contradicting each other. In reality, we must compromise. The type of project determines the scope of the planning system. The following paragraphs describe a planning system for a railway project. This is especially demanding because it has to integrate infrastructure, signaling and communication, rolling stock, and users.

A project execution plan describes the planning system with all its different plans organized in categories and with their main interactions. This is the plan on top of the planning hierarchy. It has the purpose to ensure consistency and provide an overview with mechanisms for finding compromises based on a clear hierarchy of goals.

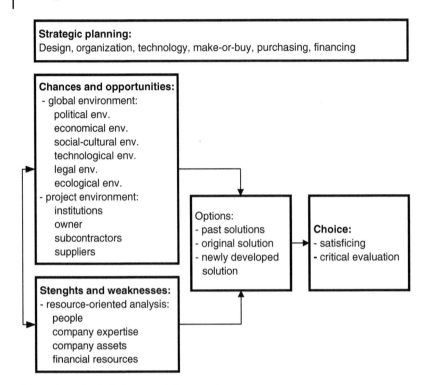

Figure 9.5 Analytic framework for planning.

Categories of the project execution plan can be corporate affairs, legal affairs, program or project management, design management, construction management, commercial management, and administrative management (Figure 9.6).

9.1.4.1   Corporate Governance Plan
A corporate governance plan regulates corporate affairs by describing general business behavior, including ethical guidelines. It also describes major roles with corresponding rights and duties.

9.1.4.2   Legal Affairs Plans
Legal affairs contain two plans. The first is a legal plan with all the legal obligations and an approach to manage them. The second is an approvals plan. Often, external checking engineers are tasked with approving all plans; sometimes, other government bureaucracies take over certain tasks of checking; and, now and then, the signature of a professional engineer suffices. These institutions can decisively affect the design and implementation processes. It is important to have clear regulations with time limits for each activity and a description of responsibilities.

9.1.4.3   Project and Quality Management Plans
The bulk of plans belong to project management. The project management plan combines these plans, similar to a project execution plan a level above. It is important to detail the connections between the different plans of project management. A quality

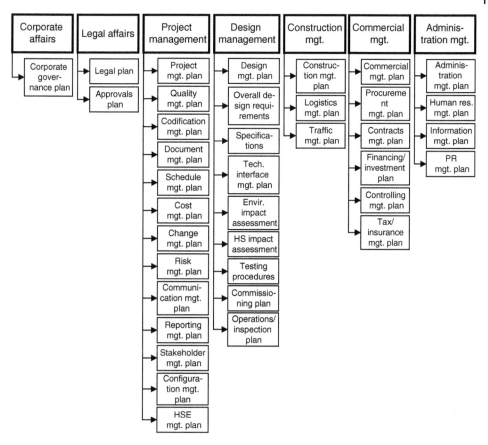

Figure 9.6 Summary of plans for megaprojects.

management plan describes the processes for achieving the stipulated quality and continuous improvement processes with root cause analysis and corrective and preventive actions. To this purpose, formal audits are required. An important part is the treatment of defects that always will occur. A zero-defect policy is a good goal but also one we can never achieve. If we try to enforce such a policy strictly with reprimands, it will be stifling.

9.1.4.4 Codification Management Plan

In a work breakdown structure, we can identify all components of the project. These can be physical items like a wall or soft attributes like customer satisfaction. Physical items have search qualities that we can measure and soft attributes have credence qualities that we can evaluate. All items must carry an assigned code for storage and retrieval. With the number of components of a megaproject, codification allows creating a functional project brain. Each code is a logical part of a system that must be consistent.

9.1.4.5 Document Management Plan

A documentation management plan for the 10 000 drawings and innumerable emails, letters, memos, and notices needs to be more detailed and specialized. Software packages help manage these documents. However, these are not off-the-shelf packages; they need tailoring. The user must define how to handle, access, and distribute documents. Rights and responsibilities must be clear.

9.1.4.6   Schedule and Cost Management Plans

A schedule management plan explains how to deal with an approved schedule. As humans cannot predict the future, all plans will be incorrect and require adjustments. The schedule management plan details how to find deviations due to planning or execution mistakes by controlling. It includes how to deal with them and how to prevent future deviations if possible. As all plans, it defines roles and responsibilities. A cost management plan does the same, only focusing on costs instead of time.

9.1.4.7   Change Management Plan

We cannot avoid deviations and must always expect changes. Therefore, a megaproject must have a change management plan. Changes occur foremost to scope, so we can also think of a change management plan as a scope management plan. When changes occur, it is important to determine who is responsible for the change, and the contract must contain paragraphs describing what to do in case of changes. This describes how to deal with changes that affect two parties, most often the owner and the contractor or the contractor and the subcontractor. There will also be changes that affect only one party, e.g. a contractor does not finish an activity on time without the involvement of the owner. The change management plan must also provide for such events by defining a process, rules, and responsibilities.

9.1.4.8   Risk Management Plan

A megaproject carries many risks. Its singularity defies prior experience, and we have to expect the unexpected. A risk management plan describes how risk management is organized. This includes a definition of who identifies risks in what way, how the risks are measured (quantitative risks) or evaluated (qualitative risks) based on a defined risk appetite, and what actions to take. Using the concept of expected monetary value, i.e. the product of probability of occurrence and the impact, it allows superposing the monetary effect of risks and profits from the cost analysis in monthly internal reporting.

9.1.4.9   Communication and Reporting Management Plans

Communication is the most important tool available to management. Accordingly, the communication management plan requires much effort. The goal is to ensure that every single stakeholder receives exactly the amount of information and attention required for optimum performance. Given the multitude of stakeholders in a megaproject and the possible number of communication channels, this is an elusive goal. However, we should try our best to get as close as possible. The main task is to define how we channel information flow through the project, again with clearly defined roles and responsibilities. This includes a definition of what information superiors give to a person and what that person must fetch himself. Important is also clarity about the use of information and communication technology. A project intranet can help organize communication. The reporting management plan defines content, distribution, and scheduling of reports. Monthly reporting is typical.

9.1.4.10 Stakeholder Management Plan
In a stakeholder management plan, we define project stakeholders and describe their expectations. Then, we must determine how to fulfill the expectations of the different stakeholders. Conflicting interests require a hierarchy of legitimate stakeholder interests. It will not be possible to make every stakeholder happy. Mitigating measures for dealing with unsatisfied stakeholders are part of the plan.

9.1.4.11 Configuration Management Plan
The contracts define the initial configuration of a project. The dynamics of the environment often cause changes to the configuration. Dramatic ones include termination of the existing construction contract and reward to a different contractor or bankruptcy of the owner. Configurational changes take place at a higher level than typical scope changes and, therefore, require different planning. A milestone trend analysis is a possible tool for configuration management. When milestones are slipping, the causes are often configurational.

9.1.4.12 HSE Management Plan
Requirements for health, safety, and the environment (HSE) are different among countries owing to different values, laws, and norms. An HSE management plan must take into account the specific conditions of a project. The goal is naturally to limit impacts on health, safety, and the environment. Ideally, we can avoid all influences.

9.1.4.13 Design Management and Overall Design Requirements Plan
The design management plan integrates all other plans dealing with design, focusing again on interlinks between plans. The overall design requirements describe which standards shall be used. Standards can be highly political. For example, there are different high-speed trains available for purchasing from different manufacturers and countries. Among other criteria, the minimum curve radii are different. By choosing a certain radius, designers can influence purchasing with or without purpose. Specifications designate product qualities.

9.1.4.14 Technical Interface and EIA Management Plans
Technical interface management plans integrate different components of a technical system. Taking a road as an example, the components might be infrastructure, traffic control, or toll collection. All components must match. Environmental as well as health and safety impact assessments (EIA / HSEA) analyze the probable impact of the design. If the impacts are negative, the plans must delineate mitigation measures.

9.1.4.15 Testing Procedures, Commissioning, and Operations/Inspection Plans
Testing procedures need to be defined because they define the benchmarks for implementation. They differ wildly depending on the project, ranging from simple in the case of

buildings to highly complicated in the case of railroads, airports, or processing plants. The commissioning plan combines the different testing procedures into a process. Commissioning might take a year to complete in the case of an urban metro system. Operations and inspection plans detail the use and maintenance of a structure.

9.1.4.16 Construction, Logistics, and Traffic Management Plans

Construction involves the largest amounts of resources, including expenditures. Billions of dollars, thousands or workers, and hundreds of managers and engineers require planned coordination. The logistics plan regulates the flow of resources in the supply chain. Every nail must find its way through this chain to the place of use. Many projects are in the middle of metropolitan areas and influence traffic flow heavily. The owner for the reconstruction of the center of Berlin after the unification in 1990 built special roads, bridges, rail lines, and two logistic centers to keep construction logistics and traffic separate. Construction of infrastructure in a city center might double or even triple the cost in comparison to construction in the green fields.

9.1.4.17 Commercial, Procurement, Contract, Financing, Controlling, and Tax/Insurance Management Plans

The commercial management plan serves integrative purposes. Procurement can amount to 90% of the total project costs if the contractor chooses to subcontract all works and is happy with 10% overhead. The procurement plan must contain a strategy and a description of the purchasing process defining roles and responsibilities. All purchases require contracts and it is the contract plan that defines roles and processes with regard to drafting and administrating contracts. Cash flow is of paramount importance to any economic endeavor and is the difference between incoming and outgoing money. Financing describes incoming monetary resources over time, investments and other expenditures outgoing payments. The contractor must control all incoming and outgoing money. A controlling plan must describe how to achieve this. Taxes and insurances also affect the cash flow and contractors must consider them alongside financing, investing, expenditures, and deviations from plans (controlling).

9.1.4.18 Administration Management Plans

The administration management plan coordinates human resource management, internal information management, and external public relations management.

9.2 Controlling

Planning determines goals for the future. When this future becomes present, we need to check whether we have achieved the goals. This is what we call controlling. However, controlling encompasses more than just comparing planned versus actual values. If the values coincide, the comparison is all we need. Due to the obscurity of the future, plans will always fail to some degree. At times, the future obstinately refuses to follow our plans. Thus, planned versus actual will show small or large differences. In such cases, controlling also includes determining the magnitude of the deviation, its causes, as well as corrective and preventive action. This is easy to state, but at times, difficult to achieve because sometimes, the causes for negative deviations originate from superiors in the hierarchy. Then, there might ensue a fight between improvement and power, with the winning side often being the more powerful one.

Controlling follows the activities of the Deming circle (plan, do, check, act ... and re-plan re-do, re-check, re-act). This is the core of continuous improvement processes. A company or ICJV that places power over continuous improvement will suffer the consequences. Controlling requires a company culture wherein learning and improvement are at the top of a list of values, and blaming and punishment are at the end. A certain tolerance for mistakes is indispensable. The goal is to eliminate mistakes, but the understanding must be that we cannot avoid them. Repetition of the same mistakes is inadmissible.

The goals that we formulate by plans do not necessarily complement each other. Goals can be competing, complementing, or neutral (cf. Section 5.3). The goals from each of the plans enumerated above will fall into one of these three categories. Neutral goals demand no attention but competing goals require a lot. We must transform the bundle of goals emanating from planning into a hierarchical system. Then, it becomes more important to optimize the goal system over the partial goals. We must be able to sacrifice partial goals to achieve overall success.

It also becomes clear that we must develop the planning system with regard to controlling, and some controlling requirements will influence the extent of planning. This is especially true if an ICJV develops controlling as plan fulfillment. It is also possible to limit controlling on budget control, benchmarks, or incentives. Given the complexity of megaprojects, a combination promises the best result. Budget control is fundamental, followed by control of plans, checking with benchmarks, and incentives for goal achievement. Most often, companies incentivize keeping or exceeding the planned profit margin.

Controlling can have different foci. Strategic control works like a radar, scanning the project horizon for any risks to the chosen strategy. It cannot have a clear direction and cannot compare clearly defined values with the status. It involves intuition.

Control of assumptions checks whether underlying assumptions are still valid. This requires, of course, that the planners spell out project assumptions clearly. If this is the case, a comparison between actual and planned is possible (Figure 9.7).

Figure 9.7 Controlling system.

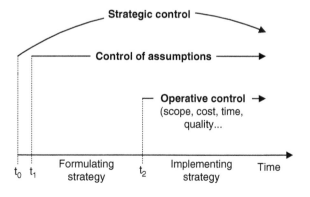

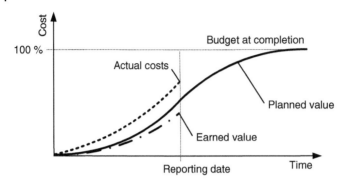

Figure 9.8 Earned value management.

Operative controls have a clear focus: we control, for example, cost, scope, and schedule with the help of earned value management. The basis for earned value management is the planned cost development over time. This typically takes the form of an S-curve (planned value) and measures construction speed (US$/month). All speed phenomena have acceleration at the beginning, most often a constant speed in the middle and deceleration at the end (Figure 9.8). The earned value equals the completed construction works multiplied by its contract price, possibly with an allowance for changes and variations. In Figure 9.8, earned value lags behind planned value, i.e. the project has a delay. Unfortunately, actual costs (as accrued from external and internal invoices) are also higher than planned. At the reporting date, the project is behind schedule and above planned costs. This is alarming but, unfortunately, happens quite often.

I have described controlling as a problem of comparing actual versus planned values with a considerable number of different goals which need controlling. From this perspective, controlling is also a tool for coordinating divergent efforts. Three theories can explain such coordination: feedback control, clan culture, and principal/agent theory. Feedback control has a clear focus. A heating or cooling system can illustrate the mechanism. We set a temperature (planned value) and the controlling mechanism checks the actual temperature at intervals. If the deviation exceeds a certain tolerance, cooling or heating begins. An isolated schedule control follows the same principal. Every month, we compare the actual and planned completion of activities. If there is a deviation beyond the allowed tolerances, re-planning of the remaining activities starts, with the goal to keep the finishing date.

Clan culture does not require coordination. In a clan, everybody knows his place and understands what to do. Only external influences need controlling. The principal/agent theory allows for different interests within a group to influence outcomes. It can serve as an analytical tool to set up controlling with regard to internally divergent interests and by setting appropriate incentives. Typically, in a megaproject, we find controlling mechanisms based on feedback controls and the principal/agent theory.

Figure 9.9 shows a cognitive map of participants in megaprojects concerning controlling. Cost, time, and quality control are of the highest importance. It is clear that a focus on this "magic triangle" holds risks. If, for example, the safety or environmental standards fall behind, the owner or authorities might stop the project.

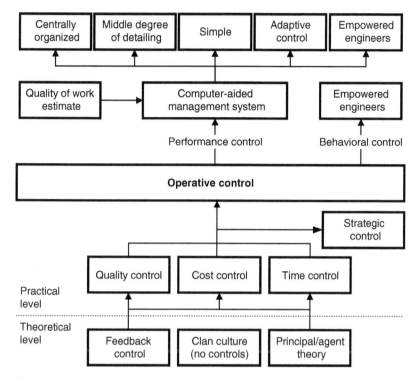

Figure 9.9 Cognitive map for project controlling.

 Safety as a risk = (obs.) A large international contractor acquired a segmental bridge project in the USA. The American subsidiary refused help from within the company for handling the overhead girders bought from another company project in Asia. Quality, cost, and time were under control, but one of the girders toppled, fell of the piers, and killed three people. The owner stopped the project.

Strategic control becomes the specific task of the top management and requires attention by everybody else, checking the horizon for risks to the plans. Operative control has a sharper focus. It is split between control of behavior and performance. Managers exert behavioral control through communication, but it must be clear what behavior is acceptable; the ICJV needs to set norms. The threshold of acceptable behavior depends on country culture, company culture, and the governance rules of the ICJV.

 Bribes = (obs.) How a society handles bribes is a question that depends on history and is part of the culture. In Prussia, the kings modernized the civil service in the eighteenth century by providing the servants with sufficient funding.

This became a German tradition. In Asian countries, civil servants had to rely on bribes or tea money for a decent income (of course, at times, it became indecent). This continues today. In a Thai/German ICJV, it was acceptable for Thai managers to complement their low company incomes with tea money. The same was not acceptable for expats with high income levels. Such different behavioral standards make behavioral controls a nightmare. Transparency International provides data on bribes.

(corruption perceptions index, https://www.transparency.org)

Performance control relies on computer-aided management systems. The input data come from planning. Given the importance of profits for contractors, budget control is of paramount importance. Thus, the quality of the work estimates for establishing budgets is significant. For every deviation, there are two basic questions: (i) Is the plan correct? (ii) Is the execution correct? Since estimating is fraught with uncertainties, as the name already suggests, planned values are often wrong. At other times, execution is suboptimal.

Most of the actual data come from the construction site, and this prompts the question whether construction management should be in charge of controlling. The principal/agent theory tells us that this is not a good idea as construction management might want to hide some problems. For this reason, the majority of the interviewees want project management to organize controlling. However, engineers working on-site (construction management) should collect the progress data. Such a split between the responsibilities for data collection and data analysis serves the ideas of the principal/agent theory well and allows for the specialization of those best qualified.

The controlling system should be as simple as the complexity of the data allows. Oversophisticated data management systems have a tendency to obscure more than they reveal. Whether a system is simple or not depends on the level of exposure of those handling the system. Learning processes make the application of more demanding software simple. This makes it preferable to stick with one system from project to project instead of always installing the newest software. An analysis of the requirements helps find a "simple" software that serves the purposes.

The level of detail should range in the middle of a continuum. Too general an approach might not divulge some important problems; too much detail might obscure problems. When the data reveal a problem, it should be possible to find the source in order to develop a solution. Should the profits slip, it should be possible to locate the problem on a lower level (labor, material, equipment, subcontractors), and then again determine a lower level as a source (e.g. what kind of material is too expensive). Once we find the original source of the problem, preventive action is required. For example, if a sealant is too expensive, the plans must show what the estimate provided as a budget for this sealant. This is a middle degree of detailing.

As mentioned, plans can be wrong, and they will be wrong at times. There is a debate over whether we should correct the plans in such a case; we call this adaptive controlling. Some argue that it obfuscates the original benchmark for the overall project success; others wonder why a wrong benchmark should apply. The whole purpose of controlling is to know the right direction and to make sure to follow this direction. Therefore, managers and engineers voted for adaptive controls.

9.3 Organizing

As planning and controlling are twin functions, so are organizing and directing. Organizing creates structures and rules that regulate future decision-making and action. Directing is present decision-making and action. As such, directing has the advantage of finding specific solutions, while we develop general solutions by organizing. In megaprojects, many problems occur repeatedly, and it makes sense to solve them through organization. We cannot foresee all future problems, and some occur only once or twice, and so directing provides a better approach. The more people work for one project, the more efficient the organizational rules will be. Megaprojects start with a small initial group and employ thousands at a later stage. Quite naturally, the focus will shift from directing to organizing during different stages of an ICJV. Also, quite naturally, there exist neither specific rules nor structures for the project at its inception. We would have to transfer all these from other projects. This is not a good approach for megaprojects. We need to develop organizational solutions for the specific conditions of each megaproject.

National cultures differ in two important ways, i.e. with regard to power distance and when dealing with the future (uncertainty avoidance). Hofstede (1984) introduced these two terms based on research at IBM. Power distance measures the differences in behavior between superiors and subordinates insomuch as the subordinates accept it. If power distance is high, superiors and subordinates differ largely in their accepted behavior. Uncertainty avoidance describes how different peoples deal with the future. Peoples with a high uncertainty avoidance worry greatly about tomorrow, while those with low uncertainty avoidance worry to a much lesser degree. I introduced these topics in Section 7.1.5 and will discuss them in detail in Chapter 12, which focuses on cultural management. Mintzberg (1980) also found five different coordination mechanisms for an organization through a literature review:

1. Direct supervision through hierarchy
2. Standardization of work processes through a complete bureaucracy
3. Standardization of work outputs through a divisionalized structure
4. Standardization of skills through a professional bureaucracy
5. Mutual adjustment through an adhocracy

Hofstede and Hofstede (2005) combined Mintzberg's five basic organizational structures with the cultural dimensions of power distance and uncertainty avoidance. The idea behind Figure 9.10 is that different countries score differently on the power distance index (PDI) and the uncertainty avoidance index (UAI) and that they prefer appropriate organizational structures.

The United Kingdom (low PDI; low UAI) prefers mutual adjustments in form of an adhocracy: qualified people join together to solve a problem, then disperse again.

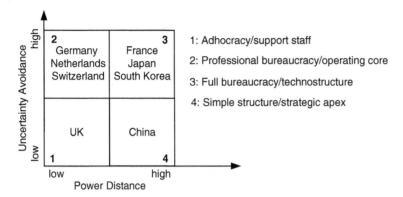

Figure 9.10 Culture and preferred organizational structure.

Germany, the Netherlands, and Switzerland (low PDI; high UAI) like to invest in the future by standardizing skills. Workers undergo a 3-year apprenticeship. What they learn during such an apprenticeship requires no supervision later. The resulting organization relies heavily on the workers (professional bureaucracy). The preference for a low PDI places workers and managers on a similar level. France, Japan, and South Korea belong to the group of countries with a high PDI and high UAI, with a tendency toward a full bureaucracy based on rules. In such an organization, the hierarchical position defines power through the allocation of rights and responsibilities. China and Hong Kong (high PDI; low UAI) have an inclination toward a simple hierarchy with much power vested at the top. The USA shows middle values for PDI and UAI, leaning toward a reliance on the middle management and standardized work outputs.

It is amusing to find these thoughts imprinted on business cards. Business cards of British managers always mention the associations of which the manager is a member. An adhocracy relies on these networks. German business cards seldom fail to indicate the educational qualifications, and French ones show the position in the hierarchy.

Defining characteristics of megaprojects are complexity and an extremely high density of tasks at the beginning. A simple hierarchy will fail because the top manager will act as a bottleneck, slowing down decisions. A full bureaucracy has to rely on rules from other projects, which will not be optimal; some rules even might prove to be outright obstructive. Much better prepared for megaprojects are people with a tendency for adhocracies, which is just another word for a project organization or for professional bureaucracies, which rely on a broad basis for problem solving. Especially at the beginning of a megaproject, delegation and flat hierarchies are not an option but a necessity.

 Preferred organization = (obs.) The owner of the Taiwan High-Speed Railway Project awarded contracts to a number of ICJVs. There were, for instance, two Japanese/Taiwanese, three South Korean/Taiwanese, and two German/Taiwanese ICJVs. The Germans introduced flat hierarchies with delegation of power based on their organizational tendencies as well as understanding of megaprojects. The two Japanese-led ICJVs refrained from establishing a full bureaucracy based on megaproject experience. The same held true for two of the Korean-led ICJVs. The last one had the smallest lot (while still being a megaproject) and the manager organized it as a simple structure. He understood himself to be a general leading his workers into the production battle. This ICJV had the worst results.

The most important tasks in organizing are: (i) creating an organizational structure with the help of an organization chart and based on a work breakdown structure; (ii) creating a process structure, also based on a work breakdown structure; and (iii) organizational rules.

9.3.1 Organizational Structure

We depict organizational structures by organization charts. This is the task of the project manager, and he must publish and explain the chart. It will change with time as the organization grows from a nucleus to full size. Every publication of an organization chart will entail discussions, as project members will also see it as a way to depict power: Who is on the rise and who is on the fall? A project manager does well to help the sensemaking process. As the organization increases in size, managers will have to hand over responsibilities to others. This does not mean that they have done a bad job or lost power.

Power = (quote) "Organizations generate power; it is the inescapable accompaniment of the production of goods and services; it comes in many forms from many sources; it is contested; and it is certainly used."

Perrow (1986, p. 265)

The first step in creating an organizational structure is a work breakdown schedule. Creating a work breakdown structure is the analytical process of breaking down the whole project into work packages. Functions or areas typically comprise the top level. Once we have created work packages, we can start the inverse process of integrating them into an organization structure. To this purpose, we assign work packages to positions, assemble positions into teams, create departments from teams, and place the departments under project management. Line positions are responsible for advancing the core tasks of the project, and staff positions support line managers. This is a theoretical possibility, and it works to some degree.

A parametric approach is more promising. Starting with a structure from a similar project, we can adjust the structure to the prevailing conditions. Once we have established a core structure, we can add positions as required. Instead of developing a full structure at the beginning, we develop it step-by-step, always observing the success of our previous decision-making. Again, it is better to engage in a learning process.

Most people organize projects and, accordingly, ICJVs as functional organizations. The second hierarchical level defines an organizational type – and here, in megaprojects, we

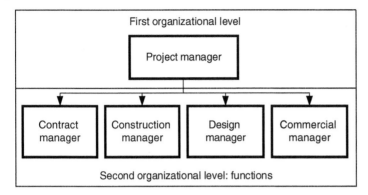

Figure 9.11 Functional organization for a metro.

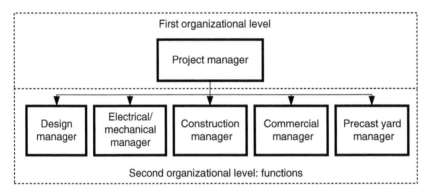

Figure 9.12 Functional organization for an expressway.

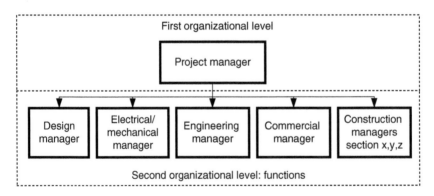

Figure 9.13 Functional organization for a rail project.

almost exclusively find functions. Figures 9.11–9.13 show three slightly different functional organizations for megaprojects.

The staff functions are not shown. In an organization as in Figure 9.11, the project manager will have support from a schedule manager, a cost manager, and a general assistant. In the following Figure 9.12, the line function "contract manager" has vanished and become

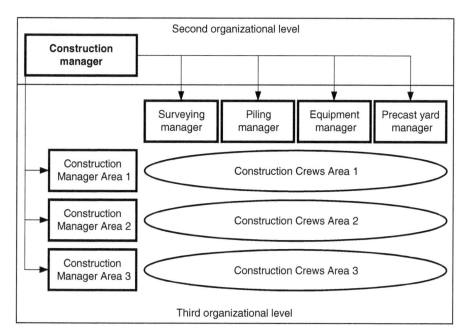

Figure 9.14   Matrix organization on the third organizational level.

a staff position (not shown). It is open to interpretation what constitutes direct work (line) or support work (staff).

Other major options would be divisional or matrix organizations. In some projects, we find a matrix organization on the third level for construction works (Figure 9.14). (Nominally, this remains a functional organization, as the second level determines the designation.) Engineers and supervisors in a specific location will have to respond to and coordinate with two different superiors on the same level, one being responsible for an area, the other for a specific task. Normally, this causes few problems in a project, and if there are any, the construction manager will have to resolve them. Matrix organizations lead to a more efficient use of specific resources.

Advantages in a functional organization are specialization and clear responsibilities. We can and must mitigate typical disadvantages, such as overall integration and overloading the project manager by delegation. Delegation means that coordination between functions is not only the task of the project manager but also that of the functional managers. This normally works well because the group of top managers (first and second hierarchical levels) is small.

The matrix organization on the third level brings efficient conflicts (!) for the use of limited resources. Area managers have a tendency to hoard resources, while specialty managers make the best use of their resources. While theory blames matrix organizations for power struggles and costly delays, it will be hard to observe this in megaprojects: the group of responsible persons is too small, and the construction managers can always solve the problem in the end.

Delegation is the organizational key to success in megaprojects. The workload is too high to be concentrated in the hands of a project manager. The distribution of the learning process and the workload is mandatory. It means that the project manager refrains from making all decisions by himself and that he must accept solutions proposed by others even if he has a better one. The guiding principle is that many roads lead to Rome. There might not be one best solution; at least, this remains unknown. This is asking a lot of the project manager because he will stay responsible for failure. The risk of failing by concentrating all decision-making on the project manager is much greater than the risk of failing by bad decisions through qualified functional managers. In the end, it is a question of balancing delegation and coordination through teamwork.

Figure 9.15 shows how the ICJV for the BangNa expressway organized delegation. Each function (administration, construction, design, MEP) had to take care of some major tasks with full responsibilities. Project management would tie everything together, as indicated by the thin gray outer ring. This ring stands for integration and adjustments if project management felt that a functional decision was not effective. This should seldom happen.

Organizations not only have a formal structure but also an informal one. While the formal structure relies on formal authority, the informal one depends on expert knowledge or personality. People who work for a company usually dispose of a network of contacts that is part of the informal organization. It would not be a good idea to prohibit such contacts. It may well be the most valuable resource at the beginning of a megaproject. If we take the necessity of delegation seriously, then we must allow people to use all resources available to them.

Laurent (1983) presents some empirical evidence on how culture affects the informal organization. Table 9.2 lists agreement with making use of an informal organization (I have transformed the data from percentage disagreement to percentage agreement). In a second

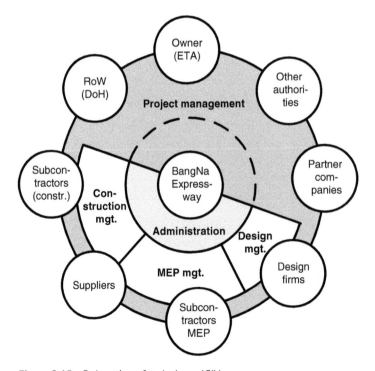

Figure 9.15 Delegation of tasks in an ICJV.

Table 9.2 Agreement to using the informal organization.

	Agreement (%)	Indulgence (%)
Sweden	78	78
United Kingdom	69	69
USA	68	68
Denmark	63	70
Netherlands	39	68
Switzerland	59	66
Belgium	58	57
France	58	48
Germany	54	40
PR China	34	24
Italy	25	30

Figure 9.16 Cultural structure.

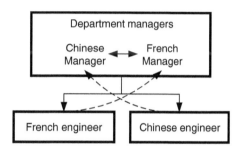

column, I have listed the corresponding country values from Hofstede for indulgence. There is a rather close correlation with the Netherlands showing a noticeable difference. It seems that indulgence has a limited positive influence on accepting an informal organization. Managers from all Western countries in the table (with the exception of Italy) find the use of informal organizations mostly positive. The value for China is an indicator that, in a simple hierarchy, this might not be true. A reason could be that informal structures undercut the power of the apex in a hierarchy.

In ICJVs, we also find the problem that a worker or engineer has a superior from another culture. If this person is not at ease speaking the common language of the ICJV (generally, this is English), he might find it difficult to communicate across these culture and language barriers. Instead, this person might ask someone from his own culture who might not have the necessary information. We are facing a cultural structure. We can solve this problem by installing two managers with different cultural backgrounds as department heads. Figure 9.16 shows the example of a Chinese/French ICJV.

9.3.2 Process Organization

Universally, we depict the process organization by using schedules. They can take the form of a Gantt chart, a network diagram, or a linear schedule (time/path diagram). The basis for all of these is a work breakdown schedule, detailing all tasks.

Schedule development is a process connecting tasks and time. When developing a schedule, engineers create several levels of detailing, from a very general to a very comprehensive schedule. It does not make sense to integrate some specific details into a general schedule; the schedule becomes unbalanced.

We can use Gantt charts best for building projects and linear schedules for projects with a linear extension, such as roads, pipelines, tunnels, and long bridges. Experts use network diagrams to develop complex schedules. For daily use, they are not very graphic as the information does not follow a timeline.

Project management must maintain control of the overall schedule for integration and controlling purposes. Supervisors as last planners have the most detailed information to produce weekly or bi-weekly schedules.

All schedules should make use of the possibility to attach resources to activities. This allows for creating a manpower curve, cash flows, and cost curves such as the S-curve (cf. Section 9.2).

Lean construction changes, among others, the scheduling approach by creating repetitive work areas (modules), standardizing work for those modules, and sequencing them into work cycles. The initial production planning takes care of all details by including last planners from the beginning and by aligning the speed of different activities through resource attribution. The planning unit is a day.

9.3.3 Organizational Rules

Rules always make sense for repetitive activities; they rely on standardization. It is much more efficient to set the start of all workdays at 07:00 a.m. than to tell 5 000 people every day when to start next day (directing). In the beginning of an ICJV, there are no rules; project management has to create them all. Rulemaking is a process that takes time and increases efficiency.

 Creating rules = (obs.) A new project manager arrived at a multi-billion-dollar rail project that was still in its beginning. He observed quite correctly that chaos and not rules governed. Then, he said something astonishing: "In 3 weeks, I will have created order." After 3 weeks, there was not the slightest change visible.

It takes many months to see the impact of rulemaking.

We define rules in project or quality management handbooks. The success of rulemaking depends on the amount and quality of organizational rules. No rules are inefficient, while too many rules are stifling; between these extremes lies the optimum number of rules. Nobody can determine the point of maximum efficiency exactly, but we can develop a feeling when enough is enough. Figure 9.17 shows a theoretical concept. It can illustrate how such a concept can help us in real life. Firstly, it clarifies a phenomenon; secondly, it prompts us to create rules; thirdly, it alerts us to the possible negative impact of rules; and finally, it forces us to combine analysis and intuition when setting rules.

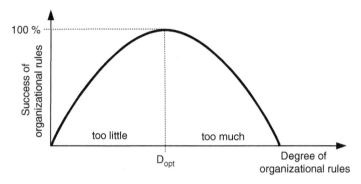

Figure 9.17 Success of rules.

Rules that find no application are certainly superfluous. However, they are worse: one day, someone will apply such a rule in an inappropriate context and wreak havoc. Maybe the following tactic might help for rulemaking: if we direct people five times to take the same action and if it is likely that we need to do it again in the future, create a rule.

Figure 9.18 presents a cognitive map of megaproject participants for organizing. They clearly separate structure and process. Engineers create organizational processes with the help of the critical path method (scheduling). There are a number of influences on the organizational structure (organization chart, rules). The ICJV structure must, to a certain degree, mirror the structure of the owner's organization. To facilitate communication, each important person in the owner's organization must find a counterpart in the ICJV. The partner companies have their own ideas about structure which are often quite divergent. Culture has also an influence, as discussed before, and so does technology (Woodward 1965). The outside influences and the decisions by the top management of the ICJV determine the organizational structure. We can depict the formal structure in an organization chart; the informal and cultural structures are hidden but also powerful.

9.4 Directing

There are different leadership theories. Best known in international construction are the "Great Man" theory and the trait theory. The "Great Man" is the born leader and he is often a self-proclaimed "Great Man." The theory denies that we all can learn to become better leaders. There is no scientific proof to the theory, and it is a highly undemocratic understanding, leading to authoritative rule. The complexity of megaprojects assures that an authoritative leadership style will fail. The "Great Man" becomes just a darned bottleneck.

Managers often use the trait theory as an explanation in the sensemaking processes of ICJVs. However, the theory does not have a scientific background. Some studies show that one trait positively contributes to success while others claim the opposite result. The only agreement in the studies is that the intelligence of the leader is never negative.

The attribution theory has much better explanatory power. People expect certain behavior, gestures, bearing, and talk from leaders. If a leader fulfills these expectations,

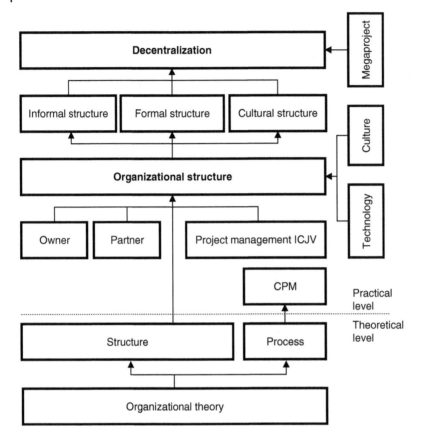

Figure 9.18 Cognitive map for organizing.

subordinates attribute leadership abilities to this person. It should be clear that the expectations depend on national, branch, and company culture.

The exchange theory postulates a relationship between leaders and subordinates, with the subordinates determining leadership status. The relationship depends on an exchange between work results, effort, and motivation on the one side and pay, respect, fairness, and development opportunities on the other. If this exchange is mutually beneficial, the subordinates affirm the leader in his position.

Finally, for our purposes, we have the contingent leadership theories from Blake and Mouton (1964) or Fiedler (1964). Fiedler considers the most effective leadership style dependent on three variables: task structure, relationship between leader and subordinates, and power position of the leader. Megaprojects are unstructured, the power position of the project manager is not established, and relationships can be good. If this is the case, relationship orientation promises the most effective leadership approach.

Blake and Mouton (Figure 9.19) proposed a managerial grid with five leadership management styles: impoverished management, country club management, middle-of-the-road management, task management, and team management. The first three are definitely not appropriate for megaprojects; two are substandard (impoverished and middle-of-the-road),

Figure 9.19 Managerial grid from Blake and Mouton.

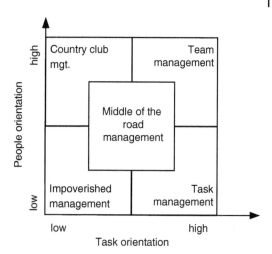

Figure 9.20 Leadership in ICJVs.

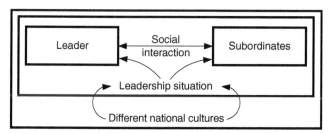

and a megaproject is definitely not a country club. The choice, therefore, remains between task and team management. The complexity of megaprojects demands a team approach. In the framework of Blake and Mouton, this stresses the importance of task orientation without neglecting a people orientation.

Whatever theory one might tend to, it is difficult not to understand leadership in ICJVs as a social interaction (exchange) where the leaders and the subordinates together determine the effectiveness of a specific approach (Figure 9.20). This approach depends on the leadership situation (contingency) and on national cultures (attribution). Interaction, contingency, and culture are the ingredients to a great variety of effective leadership in megaprojects around the world. Rare are managers who can adjust to all these different demands. We need to understand that a manager who was successful in one project might be a failure in the next. The aspect of social interaction also explains why a hierarchy is not suited for megaprojects.

 Leadership = (obs.) In many Asian countries, leaders are expected to remain calm under all circumstances, thus proving to be on top of the situation. In many Arab countries, subordinates expect a leader to prove his power by banging the table. A leader in Asia cannot behave as he does in Arab countries without damaging his position. An international leader must be able to switch between the two styles, which only the best actors among them can do well. This observation explains the influence of attribution and its dependency on culture. It also describes why some managers are very successful in one culture and fail in another.

It is not only true that social interaction, contingency, and culture are different for each megaproject, but also that megaprojects go through a life cycle from signature to handover, from a ragged beginning to a steady state, and back to the frenzy of finishing on time (cf. Section 6.4). The different demands of the activity groups might also cause a manager to succeed for some time, then fail in the same megaproject. We need to achieve a fit between demands and abilities depending on place and time.

Management failure = (quote) "Next are the potentially competent, balanced managers in perfectly doable jobs, just not the jobs for them. So they become unbalanced and therefore incompetent – misfits quite literally... Fit can also become misfit when conditions change, so that positive qualities turn into serious flaws."

Mintzberg (2009, p. 203)

Figure 9.21 summarizes the discussed influences. Organizational and national cultures comprise the first group of influences, the situation (contingency) a second, and the social interaction a third; the leadership theories held by the subordinates and their subsequent behavior, the characteristics and behavior of the leaders, the acceptance by subordinates, and the perceived effectivity by both leaders and subordinates are important. These are complex interrelations.

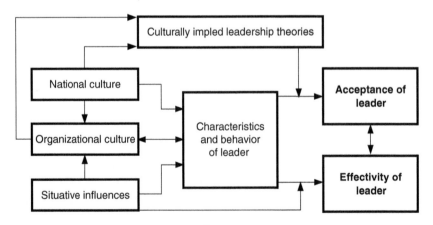

Figure 9.21 Influences on leadership effectivity.

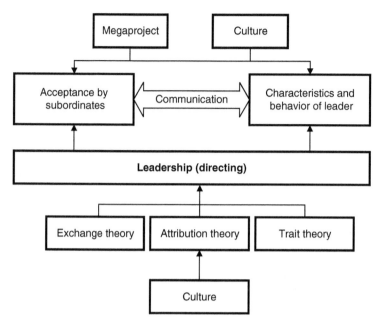

Figure 9.22 Cognitive map for directing.

The cognitive map of project participants for directing is a simplified version of the relationships in Figure 9.22. Three leadership theories have significance: exchange, attribution, and trait theory. They form the framework to understand leadership. Communication shapes the social interaction, and success depends on the quality of communication together with the expectations of the subordinates and the characteristics and behavior of the leader. The demands of megaprojects and culture influence the effectiveness of directing.

9.5 Staffing

Organizing and staffing are related. The organization chart creates positions which we need to fill with real people. Real people can manage, coordinate, communicate, make decisions, or learn. One part of staffing is more technical, which I discussed in Section 8.2.6. The other part is the social aspect of motivating and it connects with controlling and directing.

The first step in motivating is finding the right person for a position; ideally, it is the best fitting position for this person. Unfortunately, this is not always possible, but it should be one of the aims in staffing.

Herzberg's two-factor theory is observable in ICJVs. According to this theory, there are motivators that increase job satisfaction (challenge, recognition, sense of importance, etc.) and hygiene factors that lead only to dissatisfaction (income, job security, etc.). The easiest approach to motivation is increasing income. However, according to Herzberg, it does not increase satisfaction in the long run. Communication is a much better motivator, as the following quote illustrates.

Motivation = (quote) "What money? I think, talk one by one. Daytime, night-time."

Chu, Taiwanese project manager

Through communication, we can show respect and appreciation. In addition, transfer of responsibility, involvement in decision-making, and building somewhat private relationships are motivators.

Motivation = (quote) "No, I think, the motivation, you know, okay, that you share decisions. You know, you don't decide everything for your own. You motivate the people by giving them responsibility. We are talking, look, achieve this target, tell me how, what tools do you need, more equipment, more manpower, and we make a joint decision. You see them regularly, on a regular basis. You keep the relationship. You don't talk only business, you can also talk private, and I think this worked very well on this project here. Because we didn't talk only business, we had different opinions but at least we found a joint approach."

Martin, German construction manager

Figure 9.23 depicts the cognitive map for staffing. The organizational structure determines staffing and the chosen employees in turn affect the structure. Most important is the choice of the "right people", a term that is not operational. There are facts, i.e. the fit between demands of the position and the abilities and experience of the person. However, the character of this person and the cultural background seem to be just as important.

When we need to choose hundreds of employees and thousands of workers in a short time, there is little time for a thorough analysis; we must also rely on intuition.

Since the path forward at the beginning of a megaproject is not clear – I have described it as a learning process – detailed job descriptions are not helpful. At times, they can be very negative. The "right people" must have the ability to find out what they need to do; managers cannot always tell them. Initiative is important.

Along with staffing, we also must create and develop an administrative department with the task to prepare, conclude, and administer work contracts. Besides payroll and motivation, development of personnel is important. Few megaprojects last less than 3 years, and this is ample time for developing younger engineers, maybe even transforming them into managers.

Personnel development = (obs.) In an ICJV, we have two options. We can use the abilities of the personnel to a maximum by "milking" the personnel, or we can move people to other positions to help them develop. In the short run, "milking" is definitely more efficient. In the long run, it is development. A project manager who is "milking" his people will earn a negative reputation, especially with young employees. They will try everything to stay away from such a manager. A project manager who helps his employees develop will earn a positive reputation with a chance of attracting the brightest. It is better to work with bright and motivated people who are lacking a bit in experience than to work with demotivated dull people who can do one thing very well. The result of our efforts depends on skill and motivation.

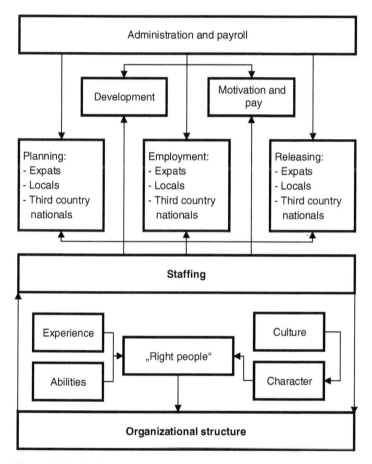

Figure 9.23 Staffing.

10

Meta-Functions

The difference between management functions and meta-functions lies in their temporal and cognitive qualities. We concentrate on management functions for a variable amount of time and in a variable order. Concentrating means that the relevant brain activities are running in the foreground. This creates the constant interruptions in the day of a manager. Meta-functions can occupy the foreground or the background of our brains, and they are continuously active; however, the strength of the brain activity varies. Mathematically, we can describe management functions as discreet (Figure 10.1) and meta-functions as continuous (Figure 10.2).

Figure 10.1 shows functions and time but not the intensity of the cognitive activities. This is close to 100% for all functions at a given time assuming full concentration on the task. Figure 10.2 shows exemplarily the intensity of the four meta-functions over time as background activities.

Figure 10.3 reminds us of the place of meta-functions in the management model.

10.1 Decision-Making

Decision-making depends on the available information and its quality. The information we deal with in megaprojects is typically complex and consists of a bundle of single information chunks; we are dealing with systemic information. This allows us to generate four different environments (certainty, uncertainty, risk, and ambiguity). The available information ranges from complete (100%) to non-existing (0%). For practical purposes, neither extreme exists; we never have complete information nor do we ever make a decision without some information. Therefore, the label "complete" means rather complete, and the label "incomplete" indicates very little information. When we have complete information, it can either be clear and easy to understand or confusing and obscure. In the first case, we face certainty, in the second, ambiguity. Incomplete information systems can contain data that allow the determination of a probability of occurrence or not. The first case constitutes a risk, and the second case, uncertainty.

Advanced Construction Project Management: The Complexity of Megaprojects,
First Edition. Christian Brockmann.
© 2021 John Wiley & Sons Ltd. Published 2021 by John Wiley & Sons Ltd.

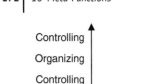

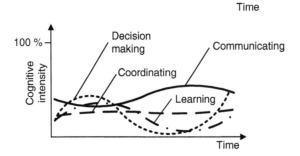

Figure 10.1 Management functions as discreet variables.

Figure 10.2 Meta-functions as continuous variables.

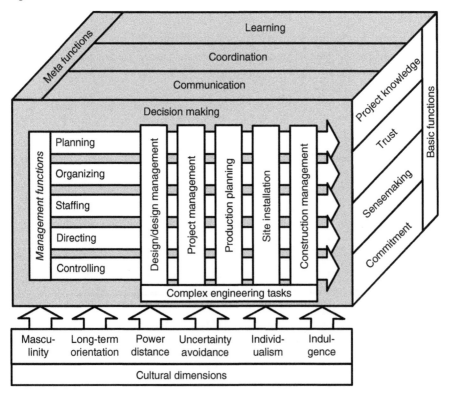

Figure 10.3 Meta-functions in the descriptive management model for megaprojects.

These considerations allow us to look at the situations that managers face when making decisions in megaprojects. A theoretical sequential process includes the following steps (Erichsen and Hammann 2005):

1. Phrasing of the decision problem
2. Determination of the goals
3. Definition of the alternatives for action
4. Definition of the environmental situation
5. Appraisal of the consequences
6. Search for a solution
7. Implementation and control

Classical normative decision theory (expectancy value, minimax, maximax, and Hurwicz or Savage/Niehans rule) assumes a sharply defined problem, knowledge of all alternatives and their consequences, and the possibility to make the very best decision. However, incomplete information about the problem and the consequences are pervasive in the practice of megaprojects. We face undefined problems, and the first step is to structure the problem in a way that includes all the important aspects of the decision-making situation. Managers need to distinguish between essential and non-essential. Next, they must determine possible outcomes and evaluate their consequences. This requires imagination and analysis. Finally, they must leap across the unknowns (decide) and accept one solution. This means finding a balance between the desirable and the achievable. All this describes the decision-making model of the New Institutional Economics (NIE) based on limited rationality and cognitive limitations (Figure 10.4).

It is no fun to make a decision based on limited information. Most managers avoid doing so, but the relentless pace of megaprojects will not allow this. Making decisions without a good amount of information will necessarily result in some bad choices, and sometimes, superiors chastise the project manager half a year later, judging on the basis of an evolved information base. Top project managers know this and try to avoid uncertainty by making decisions only when necessary with a maximum amount of information. Most of the time, a problem triggers a decision-making process. Quite often, the decision is not solving the

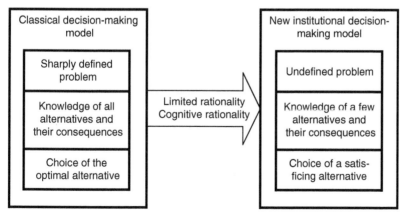

Figure 10.4 Classical and NIE decision-making models.

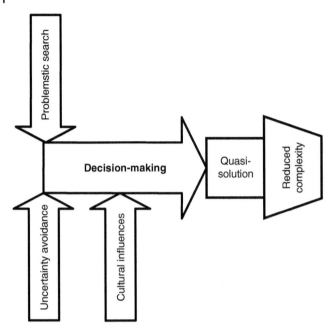

Figure 10.5 Cognitive map of decision-making.

problem completely; the result is a quasi-solution. However, even quasi-solutions reduce complexity and lead step-by-step to a full solution. Cultural impacts can arise from a more collectivistic or more individualistic approach as well as from the leadership style based on power distance. These points show up in the cognitive map for decision-making (Figure 10.5).

Cyert and March (1992) modeled the steps as shown in Figure 10.6. In this model, we find the problemistic search, uncertainty avoidance, and quasi-resolution of conflicts, as discussed earlier. In addition, we find organizational learning.

The assumptions for this model are: (i) goals act as independent constraints, (ii) the use of limited rationality, (iii) satisficing, and (iv) the sequential directing of attention. Impacts from the environment trigger the process, i.e. someone observes a phenomenon that requires action. Typically, groups and not individuals make important decisions. The observation motivates decision-making, the search is simple, and we are prone to bias. By learning, we adapt our goals, attention rules, and search rules.

As mentioned, an observation from the environment starts the decision-making. If there is certainty, we can achieve the goal with standard decision rules. If there is uncertainty, we need to find a negotiated solution between the involved parties. If we cannot reach the goal, we must change our search and decision rules. In case of success, we can add the new rules to the set of standard rules. Finally, we evaluate goal attainment. If there are no problems, attention shifts to the next problem.

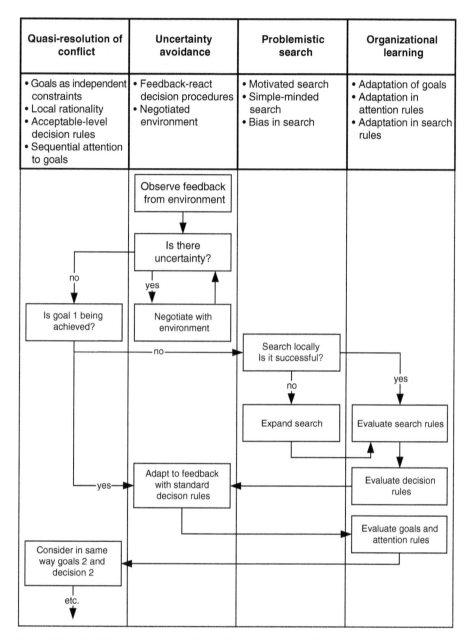

Figure 10.6 Decision process according to Cyert and March (1992). Source: Cyert, R., and March, J. (1992): *A Behavioral Theory of the Firm*. Malden, Blackwell. Reproduced with permission of John Wiley & Sons.

There is a large difference between the cognitive map of Figure 10.5 and the model by Cyert and March. The cognitive map is much simpler. However, it contains the most important components of the model. The following quote supports the cognitive map of Figure 10.5 by referring to an impact (problemistic search, someone has advances an idea) and standard decision rules (guidelines of the joint venture).

Feedback from environment = (quote) "Yes, because, I think even in the management, these managers had a different task to fulfill, so somebody, he has to bring the idea and then the decision could have been taken according, I think, to the guidelines of the joint venture."

Giovanni, Italian construction manager

Negotiated feedback can refer to facts or opinions from other people. Project managers seldom make important decisions alone.

Negotiated decision = (quote) "I mean, the important decisions were made, you and the project manager discuss the decision which has to be made with the construction manager and then with his co-project manager and then they made, how you call it, then they seek the approval of the joint venture board and this was it."

Michael, German design manager

Decision-making in megaprojects is an iterative process, as shown by Cyert and March. It involves different people at different times and additional information that becomes available in the process. The process takes time.

Iterative process = (quote) "Typical decision process. It needs time; it needs time because it is a joint venture and two partners, two different cultures, so it takes a little bit more time. You need at least two or three circles where you have to discuss and you find out more details because you understand more, you can think more. Then you go into the next meeting and discuss with the partners, try to find out more background information which was not given – not on purpose, but because maybe they were not important from the other point of view. It just needs more time."

Erik, German project manager

The following observation illustrates the decision-making process in an ICJV, and the relevance of the theoretical model of Cyert and March becomes obvious.

Decision-making process = (obs.) The BangNa Expressway is a highly innovative project (cf. Section 13.1). The overall design required a decision on inclined elastomeric bearings. This was a request from the design. Since neither the designer nor the contractor had any experience with inclined elastomeric bearings, uncertainty existed. The behavior of such bearings under inclined loadings were unknown. It followed the negotiation with the environment: a search for information on inclined elastomeric bearings. There was no information available, no codes anywhere in the world, no test results, and no research. This search was first local in the domain of the designer and the contractor; then, the two parties expanded the search to the global level. Since there was no result, the search rules changed, and the designer and contractor considered testing the elastomeric bearings. It became clear that there was insufficient time for testing. Accordingly, the contractor changed the decision rules from fact-based to intuition-based. In a final meeting with everyone present, all pros and cons were once again considered, and the project manager made a decision. This consisted of limiting the bearing capacity from the standard horizontal placing, i.e. he asked the designer to over-dimension the bearings. This procedure can also explain the cognitive map of Figure 10.5. Decision-making began with the need for inclined elastomeric bearings (problemistic search). The process took several weeks to conclude. The contractor delayed the decision until the final meeting, hoping to find more information (uncertainty avoidance). Cultural influences played no role; the designer from the USA and the contractor from Germany took a similar approach. The result was a quasi-solution. Some doubt prevailed over whether the actual behavior would conform to the expectations. The owner demanded a prolonged 20-year warranty. The contractor provided the warranty. At the time, the decision solved the problem. Only 20 years later, at the end of the warranty, did the solution become final.

The image of a captain's absolute power to make decisions onboard a ship does not apply to construction. A ship might come into acute danger, where survival depends on decisions taken in a very short time. This is almost never the case with ICJVs. There is time pressure, but one that, in general, allows some time for decision-making. Also incorrect is the idea of the project manager who makes decisions by himself as a heroic leader. Project managers who entertain such models of behavior will most likely fail. The role of the project manager is more one of a monitor-evaluator who combines ideas, opinions, and facts into a coherent solution.

10.2 Communication

Communicating is what managers do; engineers work on solutions at their desks. The latter concentrate on things, the former on people. This is even truer for megaprojects. To some degree, communication is part of all functions in the descriptive megaproject management model. A manager or engineer might elaborate a plan in his office behind a closed door. Once finished, he will have to get approval (or negotiate with the environment). This involves communication, and the plan itself is written or graphic communication. A megaproject manager who is not listening and talking, using body language or action as well as plans, rules, and presentation for communication, is not doing his job. Anecdotal evidence places the time that managers spend communicating at 70%–90% of their total time, with a larger portion of this dedicated to listening.

Communication = (quote) "Communication is not simply what managers spend a great deal of time doing but the medium through which managerial work is constituted."

Hales (1986, p. 101)

10.2.1 Megaproject Communication

Communication in construction and megaprojects depends on the structure of project organizations. Dainty et al. (2006) list seven characteristics influencing communication:

1. Uniqueness of construction projects leads to specific organizations for each project with different people and different challenges; there is no history as a base for the future of the project. Interdependence is high.
2. Construction projects tend to be awarded at short notice, leaving little time for preparation. There is no warm-up phase, only full speed.
3. Labor intensity requires a large workforce.
4. The fragmented structure of the supply chain brings together different jargons and semantics, breeding misunderstandings.
5. Reliance on a mobile workforce always throws new people together in different projects.
6. An ingrained male-dominated workforce creates communication specifics.
7. An increasingly diverse labor market with projects where different crews have to communicate through interpreters with both parties using English as a second language.

We can add to this list the complexity of megaprojects, which takes the difficulties listed here to a higher level. A specific problem is the absolute need for the delegation of authority in megaprojects, further raising the bar for communication. There are few other industries where communication is more important than in construction. Engineers tend to dominate construction projects, and very few choose such a career because of an affinity toward language, which expands the list of difficulties.

10.2.2 Communication Models

There exist different models for understanding communication. As always in such cases, this allows us to look at a phenomenon from different angles. The dialog-based model by Watzlawick et al. (1967), the encoder/decoder model by Shannon and Weaver (1949), and the four-aspect model by Schulz von Thun (1981) are important for megaprojects. The encoder/decoder or sender/receiver model is best known in megaprojects.

10.2.2.1 Dialog-Based Model by Watzlawick

Watzlawick et al. (1967) formulated a communication model based on dialog by positing five axioms of communication:

1. (Cannot not): One cannot not communicate.
2. (Content/relationship): Every communication has a content and relationship aspect, such that the latter classifies the former and is therefore a meta-communication.
3. (Punctuation): The nature of a relationship is dependent on the punctuation of the partners' communication procedures.
4. (Digital/analogic): Human communication involves both digital and analogic modalities.
5. (Symmetric/complementary): Inter-human communication procedures are either symmetric or complementary, depending on whether differences or parity defines the relationship between partners.

(1) It is impossible not to communicate because we always send messages through body language. The bored and distracted member in a meeting sends a clear message with his silence. (2) Since communication is highly personal, it requires a foundation in a positive relationship. If this foundation is solid, communication will flow much easier, even in case of disagreement. The typical attempt by managers with an engineering background to calm and control a meeting by "just sticking to the facts" is futile. (3) Punctuation sets the start for an argument, and where we start in a continuous relationship determines the outcome. This can result in a blame game where two parties go further and further back into the past to find fault with the other. As many relationships have quite a long past, it becomes an unending regress. In the end, communication is circular, and a circle has no natural beginning. Blame games are also futile. Any setting of a start is arbitrary and can be misused. (4) We typically use digital communication to transmit content and analog communication for relationships. Talking with each other is digital, but body language, voice modulation, and context are analogous. (5) Differences in relationships spawn complementary communication. The communication between superior and subordinate as roles defined by the organization chart is complementary. The communication between equals is symmetrical. As two people fill different roles at times, communication can shift between complementary and symmetrical. It might be complementary during working hours and symmetrical later over a beer.

10.2.2.2 Encoder/Decoder Model by Shannon and Weaver

Shannon and Weaver (1949) were the first to introduce an encoder/decoder model of communication. This model pays little attention to the relationship aspect, and the authors see it as a mathematical model. Figure 10.7 shows a graph of this model with the addition of communication channels and media. The sender has a certain concept in mind, which he wants another person to understand. He must first formulate the content (encoding) and then speak, write, or draw up this content for transmission. We call this feedforward, and its quality depends not only on concept/encoding/transmission but also on the chosen

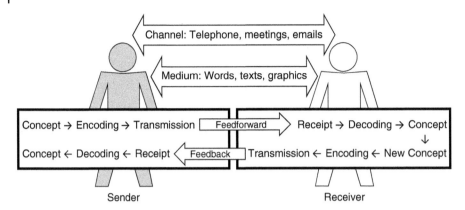

Figure 10.7 Encoder/decoder model of communication.

channels and media. The receiver must pay enough attention to shift the information into short-term memory (receipt). Then, he must decipher the words and the meaning (decoding). Seldom is a brain so empty that the communicated model will not meet some relevant information in the receiver (concept), who will then combine the two sets for information (new concept). Without the following step of feedback, the sender will never know what has happened in the brain of the receiver. The receiver now reverses the process and becomes the sender. In the end, the first sender can compare his intentions with the understanding of the receiver.

Unfortunately, much of our organizational communication contains little feedback. Not shown in this model are outside barriers to communication. Shannon and Weaver call these "noise," referring to tuning into a radio station with a weak signal. Using the model, we can well identify internal and external communication barriers and try to avoid them in our communication.

Figure 10.8 lists some typical barriers that follow from a context analysis of the construction industry and megaprojects.

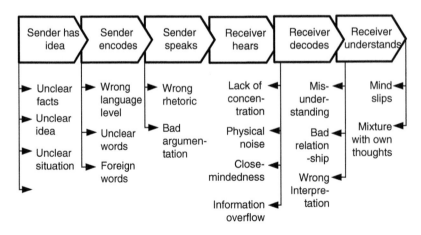

Figure 10.8 Communication barriers in megaprojects.

Idea: The sender in a megaproject faces many problems: an overwhelming work intensity that requires a machine-gun type of decision-making, no structure, and little information. He does not even have an idea of the right path forward. Accordingly, facts are unclear (ambiguous), ideas or solutions are unclear, the situation lacks analysis, and while there is generally too much information, there is too little relevant information.

Encoding: In megaprojects, the management has to address people from different backgrounds, from presidents and prime ministers to almost illiterate workers. Unclear facts, ideas, and situations often hide behind unclear wording; the ambiguity pervades everything, from concept over encoding to decoding and understanding. The highly fragmented supply chain involves many jargons and national languages. Finding understandable words is no easy task.

Transmission: An incorrect rhetoric is part of analog communication; it might confuse more than clarify. Culture also determines body language. Building an argument for different audiences is challenging and error-prone in general. It is more than that, given the diversity and pressure of megaprojects.

Receiving: Work hours are long in megaprojects – 12 hours per day and 7 days a week are normal for management in the beginning. Crews work in 12-hour shifts, and in Asian countries, the workers might even wander off to a second job. Concentration is a challenge at least for part of the time. Noise from a variety of sources interferes with the transmission (Figure 10.9). National and professional cultures provide mindsets that might predetermine the reception of new ideas (close-mindedness). Information overflow can completely overwhelm people in megaprojects so that they simply switch off.

Decoding: Misunderstandings occur when communication deals with complicated ideas and facts. Complicatedness is part of complexity; complexity in turn defines megaprojects. Accordingly, misunderstandings will occur often. The pressure of megaprojects does not always allow enough time for building relationships. The team building at the beginning

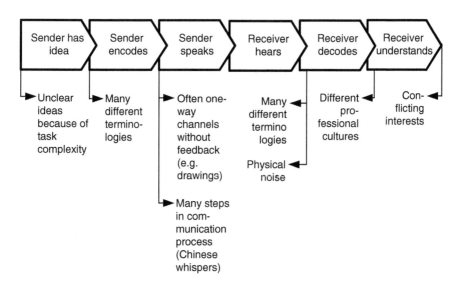

Figure 10.9 Noise in megaproject communication.

of projects is during its stormy period prone to fighting. Differently wired mindsets trigger wrong interpretations.

Action: Mind slips are again due to long work hours and information overflow. If the receiver is not able to combine and integrate the information consistently with his own ideas, an unproductive mix might be the result. Not everybody can easily give up long-held concepts for better ones. This is a time-consuming process.

Complexity causes much noise by confusing people's concepts. According to the Bible, when a united human race tried to build the tower of Babel, tall enough to reach heaven, God punished the people in that megaproject by muddling their languages, so that they could no longer understand each other. The megaproject failed, and we are still dealing with the problem today.

Much communication is predominantly one-way, from drawings over rules to directions. The extended supply chain means that information passes from one end of the world to another; Chinese whispers is a metaphor indicating an accumulation of communication errors.

Physical noise is pervasive not only on-site but also often in offices close to the site. Different cultures with their associated mindsets definitely make it difficult at times to form a common understanding. Conflicts of interest are manifold; the most important one is that between owner and contractor. The owner understands when signing the contract that he will get everything he wants for almost no money; the contractor believes the inverse: he has to provide little for a lot of money. The conflict is intractable.

10.2.2.3 Four-Aspect Model by Schulz von Thun

Schulz von Thun (1981) presents a behavior-based model containing four aspects of information:

1. Facts
2. Self-revelation
3. Relationship
4. Appeal

The model helps analyze communication behavior and detect barriers. Every sentence contains some statement (which might be correct or incorrect). It also reveals something about the speaker, such as preferences or feelings. It is a statement of the status in a relationship or an attempt to change the relationship. It also includes an appeal to the receiver.

The sentence "Please help me resolve this difficult problem" states a fact, i.e. there is a problem. It also reveals that the speaker cannot find a solution alone. The word "re-solve" (instead of "solve") means that the problem is recurring and that the speaker has failed repeatedly to find a solution. He shows himself as weak, which he can only do in a positive relationship. The appeal is very clear: "Help me!"

10.2.3 Communication Methods

Managers use different communication methods. In a study of R&D projects, Kyriazis (2007) found the following order of preference (Table 10.1).

R&D projects are similar to construction megaprojects as they start from scratch and create innovation. Differences exist in size and duration. Megaprojects have much more

Table 10.1 Communication methods used in R&D projects.

	Rank
E-mail	1
Impromptu face-to-face conversations	2
Scheduled one-to-one meetings (face-to-face)	3
Impromptu one-to-one phone conversations	4
Reports	5
Scheduled one-to-one phone conversations	6
Voice mail	7
Informal face-to-face conversations in a non-work setting	8
Teleconferencing	9
Handwritten memos	10
Fax machine	11

available resources but have to finish at a defined point in time. Given the similarities and differences, we can expect megaprojects to have similar preferences for the use of communication methods. With the exception of reports, the first seven methods use a one-to-one communication format. Here, communication is very private and affected by personality.

Four pairs of personal characteristics especially affect communication. These pairs define a continuum with two endpoints. Most people display characteristics somewhere between the endpoints. The first pair is extroversion versus introversion, with extroversion facilitating communication while not always raising its quality level. The second pair is sensing versus intuition. Our five senses, particularly eyes and ears, define sensing behavior, and sensing rests on verifiable facts. Intuition makes fact-checking more difficult; it is indispensable in megaprojects as facts are often not available, but it does not help communication. Thinking versus feeling describes an analytical or a holistic approach. We can explain an analysis better than a holistic idea. The last pair is judging versus perceiving. Some people have a tendency to judge rather early, whereas others wait longer before arriving at a conclusion. Culture influences this strongly. In Western countries, there is a clear tendency to encourage children to judge; Eastern cultures focus on strengthening perception.

 Personality in communication = (quote) "It always comes down to individuals. How I do this, how I come across, how they receive you, how they do it, how they come across, how I receive them. Every time it's new. It comes down to cultures, to people, to communication skills. But one prerogative is there: you need to talk. If you don't communicate well, show a certain openness, and perform in accordance to what you say you were going to do, nobody will ever trust you in the long term."

Maximillian, German construction manager

With all the different communication methods available, face-to-face contact remains a preference. Face-to-face contact provides analog as well as digital feedback, builds trust, and motivates.

Face-to-face communication = (quote) "But what I'm thinking right, if some problems arise, then we have to talk face-to-face... I mean meet the schedule – that's my purpose. So I don't care, I really like to talk face-to-face and try to solve this kind of problem."

Hsu, Taiwanese project manager

Communication methods can be very culture-specific. An example is morning exercise, which the interview in the following quote details.

Morning exercise = (quote) Answers by Kim, a Korean project manager

Question:	Do you have a meeting every day?
Answer:	Every day in the early morning – the morning exercise. Do you understand the morning exercise?
Question:	No. Who is doing the exercise when you have this morning meeting?
Answer:	Before the meeting, we do the morning exercise.
Question:	How? The workers? They do a morning exercise?
Answer:	Yes. All workers and all staff.
Question:	Together?
Answer	Yes. Together in the same place.
Question	What do the Thai say?
Answer	Thai say that the first time, they didn't know about the morning exercise. But after 1 week, they follow us.
Question	So they see no problem, they just go along and...
Answer	Yes. They see no problem.
Question	Because, you know, I just saw a movie, an American movie, called *Gung Ho*. It's about an American going to Japan. There is something like the morning exercise. The American looks at it and makes fun of it. On the other hand, everybody in America tries to do some exercise. So why not do it together? What's so funny about it? I don't know. But it's very different, the approach, because what you do is, you get everybody together in the morning – workers, staff – everybody sees each other, and on the European site, everybody just goes his way.

Question	Yes. In my opinion, that it is quite a culture thing. In the morning exercise, I give my workers release of the body, and after the morning exercise, the site engineer said to the workers to ensure their safety, checking their safety helmets, safety jackets, and shoes. If they forget the safety helmet and shoes, we send them back.
Question	Now, when you think of this morning exercise, when you do something together, there is some bonding.
Answer	Yes, some bonding.
Question	Do you do these exercises also for team building, team spirit?
Answer	Yes.

Asian project manager giving the answers

10.2.4 Communication Organization

In construction projects, especially in megaprojects, we need to organize communication. With 500 staff and 5000 workers as workforce, we face $r = n\,(n-1)/2$ possible face-to-face communication channels, i.e. a staggering number of 15 122 250 dyadic communication channels!

The organizational hierarchy can help channel the communication by transferring communication responsibilities to every manager. This way, communication passes down the organization. It is important to make sure that everybody gets all the information required to do (details) and understand (big picture) his job – not less, not more. This way, we can limit information overflow and only pass on relevant information. To achieve this is probably the most difficult task in a megaproject. It requires organizing information through analysis and evaluation, coupled with strict discipline. Meetings and events allow disseminating information more generally.

Figure 10.10 shows an example of the organization of top-level communication. The first step is a stakeholder analysis, and the second is the delegation of communication responsibilities. In this example, the project manager takes care of communication with the owner, other authorities, banks, and the ICJV (partner, ICJV-board). The commercial manager has a wide range of communication partners (personnel, general administration, purchasing, contracts with subcontractors, as well as tax authorities and insurance companies). The design manager maintains contact with all design firms and expert consultants. The E&M manager connects with E&M consultants and relevant subcontractors. The construction manager is responsible for the information flow with all construction subcontractors, material, and equipment suppliers.

This figure also illustrates how delegation of work functions. The project management abstains from using information as a power tool by sharing it with other top managers. This requires trust, coordination, and cooperation, which is not always easy or given. The

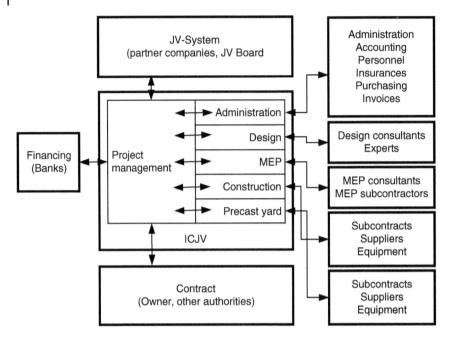

Figure 10.10 Organizing communication in a megaproject.

ICJV top management must exchange vital information so that everybody can see the whole picture.

Generally, information can flow top-down, bottom-up, or horizontal. Top-down or bottom-up communication combines different hierarchical levels, while horizontal communication remains on the same level (Figure 10.11).

Top management typically communicates goals, orders, and feedback top-down. Subordinates raise problems, report results, ask questions, and make proposals bottom-up. Horizontally, they coordinate, organize help, and give each other feedback.

Organizing communication is a task for the top management, while communicating is a task for everyone. Unfortunately, very often, a lot of information gets lost in the hierarchy.

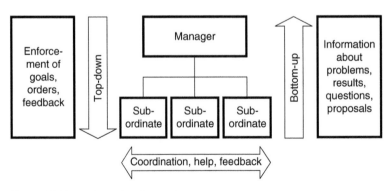

Figure 10.11 Information flow and content.

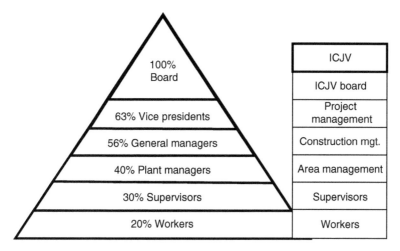

Figure 10.12   Information loss in a hierarchy.

Bateman and Snell (2002, p. 484) provide an example for information loss in an organization (Figure 10.12). As information passes from the board of directors through five levels, 80% of the information is lost. Superposing this with a typical hierarchy for an ICJV produces a loss of 70%. These are not empirical data for a construction project, but they are plausible.

The question remains: How do we organize information in an ICJV so that the responsible person has all the necessary information at every workplace? For this, planning and implementing communication according to the planning with frequent feedback are required. It is a never-ending task.

Figure 10.13 presents a cognitive map for communication as held by megaproject participants. At the beginning, they see the complexity of the project (or the chaos). This leads inevitably to the delegation of tasks, power, responsibilities, and information. They base their understanding of communication on the encoding/decoding model of communication and stress the importance of feedback.

Different types of noise make communication problematic. An important one is culture, be it national, industry, or company culture. It is difficult to distinguish between cultural and personal traits in communication. Often, we blame an individual for a behavior that is well acceptable in his own culture. Culture has an influence on our personal characteristics, and these are important in communication. The semantics of different languages (national, industry, or company) in the extended and diverse supply chain cause additional problems, and so does power distance (status). Giving no feedback highly complicates communication.

Through communication, it is possible to build trust between individuals and within groups, and it motivates individuals and groups. Misunderstanding creates cultural barriers, and the only way to remove such barriers is to discuss them. If successful, we can differentiate between personal and cultural characteristics and prepare for problems. Typically, diversity is a chance and many see it that way in megaprojects. Diversity increases the different approaches available to solve problems. This far outweighs the communication problems.

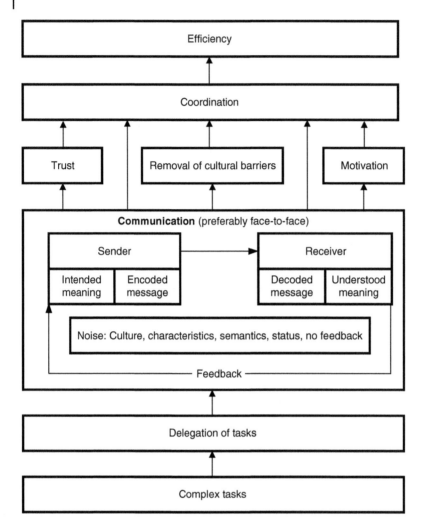

Figure 10.13   Cognitive map for communication.

Trust, cultural understanding, and motivation are of great help to coordinate a large workforce. Better coordination allows us to work more efficiently, and thus, deal better with complexity. In sum, communication helps solve one of the biggest challenges of megaprojects – its complexity.

10.3   Coordination

Imagine a sports team, be it basketball, football, or soccer, and all players moving in prefect harmony across the court or pitch. Certainly, such a team would win many games, even with inferior talent. Some teams come closer to this ideal, others less so; no team ever sustains it for an entire game. Now imagine a team of thousands of managers, engineers, and workers

Table 10.2 Coordination in orchestras and megaprojects.

	Orchestra	Megaproject
Task	Known, structured, and detailed (full score)	Known, unstructured, and general (incomplete contract)
Time	Short (hours)	Long (years)
Team	Small (tens, hundreds)	Large (thousands)
Organization	Simple hierarchy with conductor at apex	Complex hierarchy with delegation
Learning	With individual instruments, rehearsals	On the job
Understanding	Clear	Evolving
Sensemaking	Clear	Evolving
Audience, owner	Present at performance as listeners	Present throughout with the right to demand changes
Result	Harmony	???

trying the same. Even in the best cases, they will end up far from the ideal. This analogy shows how important it is to realize at least the best coordination available. In Section 10.2.4 about communication, project participants express in their cognitive map (Figure 10.13) how important communication is for coordination and solving complex tasks.

The problem of coordination exists once we start to disintegrate tasks by specializing. The division of labor is as old as capitalism and increases productivity manifold but it also requires coordination in turn (Smith 1776). The analysis of the production task is part of organizing: it leads to a work breakdown structure working from the top-down and to a hierarchy working from the bottom-up. Analysis (hierarchy) is easier than synthesis (coordination).

Sometimes, an orchestra serves as an analogy for the task of coordination. This is a misleading comparison and, in the case of megaprojects, quite wrong. Table 10.2 shows the differences.

Achieving a certain degree of coordination costs money. Having no coordination in megaprojects costs a lot of money. We face the problem of optimizing coordination.

 Coordination = (def.) Aligning plans, decisions, and action for an ICJV toward a superordinate goal.

10.3.1 Coordination Methods

We can use three different approaches to achieve overall coordination: personal coordination, technical coordination, and bureaucratic coordination (Figure 10.14).

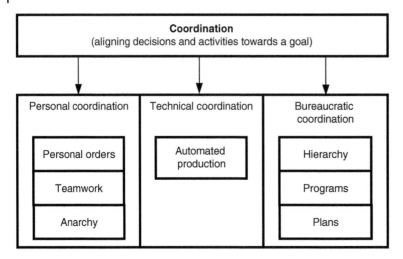

Figure 10.14 Coordination methods.

At the beginning of a project, there exist neither a full hierarchy (bureaucracy) nor plans for an automated production; personal coordination is the only available approach. Personal coordination uses communication as a tool. The initial engineering group is small and its members might not know each other. A team might not yet exist. At this time, we will find a mix of personal orders, some teamwork, and some anarchy. Anarchy designates, in this case, a voluntary cooperation without any formal structure. As the megaproject develops, anarchy recedes into the background, teams grow together, and personal orders remain. It is clear that delegation requires a large extent of teamwork.

The moving assembly line (introduced by *Ford 1913*) is the paragon of automated production. Unfortunately, it is not useful for construction. The basic idea is to move the workpiece at a constant speed and keep manpower and equipment stationary. It is not possible to move megaprojects because of their size.

The idea of lean construction inverses the process: the workpiece (structure) is stationary and manpower and equipment move through it at a constant speed. This is the production train. Modularizing, standardizing, and sequencing are the first steps toward this goal. Modules are space/task entities. We divide the whole structure into spatial units and different tasks (trades) that need to be performed in each space. Next, we standardize each task as much as possible to allow for simplification and learning by repetition. Finally, we arrange all trades in a sequence and determine the required resources to align the speed of all trades. Returning to the analogy of the production train, the modules are the stations where the train stops. The different trades are the wagons, with the most important one being the engine. Coupling between engine and wagons guarantees constant cycle times. The engine and the wagons always stop at consecutive stations.

Automated coordination requires a bit more than the described cycle time planning. The last planner approach allows integrating the participants of the supply chain by using a collaborative planning of the production train. The principle of continuous improvement helps reduce waste in the process. While the moving assembly line replaces personal communication, this is not the case with the production train. The remaining complexity of the modules and the process still requires reduced personal communication.

Bureaucratic coordination materializes in hierarchies, programs, and plans. Together with automated production, it relieves managers of the burden of communication overkill. Hierarchy defines authority through a line of command, programs provide guidelines for repetitive tasks, and plans anticipate the future and determine action for this future. As discussed, planning needs controlling and it reduces communication requirements. The same holds true for hierarchies and programs.

10.3.2 Fragmented Supply Chain

Megaprojects are truly global. Engineers on three (or more) continents might elaborate a common design connected by interfaces. Materials and equipment arrive from different countries as do members of the ICJV and subcontractors. All these people work in different time zones. Some of our newer communication tools, such as video or telephone-conferencing, that require simultaneity are of no help, but others that decouple time, such as e-mail or chats, work wonders.

Some see co-location as a solution to deal with the fragmented supply chain. All relevant parties to the supply chain have to have a front office in the same building close to the construction site. Impromptu face-to-face discussions become possible, spreading trust and motivation. For example, the owner of a large megaproject in Qatar rented a high-rise building, moved into the top two floors, provided office space for everybody else, and demanded presence.

Figure 10.15 lists further barriers to coordination. When the interdependency between departments in a functional organization is strong, coordination becomes difficult and requires much effort. This is the problem with departmental interface management. When, on the other hand, the departments are very different in their tasks and perhaps their attitudes, they have problems understanding each other. Everybody works toward a different goal set. Only coordination can solve the problem based on communication with a stress on analog communication, i.e. relationship-building.

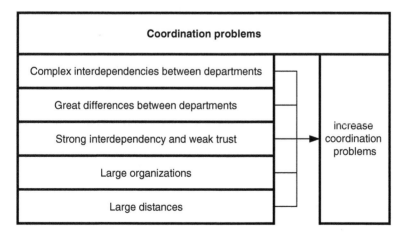

Figure 10.15 Barriers to coordination.

When there is little interdependency and little trust, pride and arrogance might be the source of gaps. Again, coordination supported by analog communication is a possible way of team-building. Large organizations need higher degrees of coordination according to Smith (1776), and megaprojects are at the edge of coordination possibilities. As discussed, we can solve the problem of large distances between departments or companies by co-location or at least very frequent face-to-face meetings.

 Coordination in megaprojects = (obs.) In a design/build road megaproject, the contractor awarded part of the design to a company in North America with work being distributed across three different offices working in three time zones, another part to a company in Europe, and a third part to a company in Asia.

The relationship with the North American company started well. They were working weekends to finish the first design package. Meetings on the job site and in North America were frequent. Unfortunately, the owner shifted the starting location to another area and pressure increased to finish this package. The same team in North America went back to working around the clock, although with less enthusiasm. The contractor pressured them in meetings to make optimistic promises; the design firm made the promises but could not keep them. Frustration on the side of the design firm and mistrust on the side of the contractor deteriorated the relationship. In the end, the contractor sent an employee to control progress in North America and to report it back. This was an open statement of mistrust and a broken relationship. It took a lot of effort through face-to-face communication to improve the relationship, and thus, coordination. The breakthrough happened when the managers responsible for the downward spiral were replaced. Communication was the only way to solve the coordination problems among the owner, contractor, and designer. Some managers were obstacles in this communication.

The observation above demonstrates the close relationship among coordination, communication, decision-making, and learning. At the center are human beings. As someone said earlier: "*It always comes down to individuals.*"

Figure 10.16 shows the cognitive map for coordination. Complexity requires decentralization and differentiation (specialization). The large number of employees in an ICJV for a megaproject allows for a large degree of specialization. The multi-organizational structure of the ICJV provides for spatial distance between project participants. The shared goal of completing the megaproject creates manifold interdependencies and the need for coordination. This coordination is especially important among owner, partner companies, the ICJV, and the supply chain. Depending on the coordination level, managers prefer different tools of coordination, such as directing, hierarchy, plans, and creation of teams. Individuals might use synchronization of actions by themselves, i.e. a form of anarchy. Higher-level means of coordination are socialization (alignment of cognitive maps), commitment, and

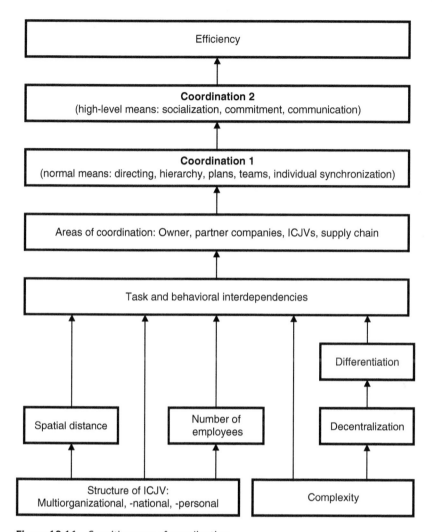

Figure 10.16 Cognitive map of coordination.

of course, communication. In the end, all these efforts will lead to a higher efficiency, i.e. less waste of resources and time.

10.4 Learning

Learning and forgetting are two of our vital brain functions. The Latin phrase "repetito est mater studiorum" shows what we have known for a long time that recurrence furthers the learning process. However, construction projects have a certain degree of singularity and megaprojects, a high degree; there is little repetition. What a contractor can learn in one project might not be applicable for 20 years in a similar project. By then, many of the knowledge bearers will have retired from the company and the remaining younger ones would have been in a subordinate position with limited access to knowledge at the time of learning. Forgetting will add to the woes.

The discontinuity of construction poses an important barrier to learning. An obvious solution is a knowledge data bank, but that is difficult because it requires the codification of implicit knowledge and it will contain many lessons learned that can never be applied again.

Learning is a very important aspect in getting successful access to a megaproject. If we think of the consequences of singularity in general and the specific singularity of megaprojects regarding the task, the organizations and people involved, the different mindsets, the required understanding, as well as the required actions, we find only one answer to solve the puzzle: we must learn how to clear the path forward.

Applying lessons learned faithfully, i.e. extrapolating a trend, will lead nowhere. Megaprojects are black swans. Lessons learned provide a starting point from which we have to depart quickly to find unique access to the focal megaproject.

Learning and innovating are interdependent. For our purposes, I will assume that learning describes a process and innovation a result. Under this assumption, all ICJV progress is part of learning. Dutton and Thomas (1984) distinguish between the source of learning (exogenous or endogenous with regard to the project or the firm) and the way of learning (autonomous or induced). Autonomous learning happens within an existing set of rules and induced learning, by changing these rules. Argyris and Schön (1978) use the terms "single loop learning" or "double loop learning" instead. For a production process, the terms "autonomous" and "induced" are appropriate because they allow placing the stimulus of learning to come from within or from outside the production process. Looking at a megaproject, learning can happen inside the ICJV or outside it (Table 10.3).

What is inside and outside an ICJV depends on contracts. Project development is typically a task for the owner, and therefore is outside. Design can take place outside (design/bid/build) or inside (design/build) the ICJV. Production typically goes on inside the ICJV. However, there are other ways of influencing the design by a contractor than design/build. If the owner decides to take advantage of early contractor involvement, then inside (contractor) and outside (owner) are connected. In sum, learning depends partially on the contractual arrangements. When devising such arrangements, the owner should pay attention to facilitate learning.

Autonomous learning happens on the job; it is a gradual process. In the end, the result of the learning – the innovation – might be very important. However, in most cases, the participants cannot detail the small steps leading to the important result. The promoters of this way of learning are often supervisors and workers. The knowledge is implicit, not codified, and hard to copy. Analysis is the basis for induced learning; managers and engineers mainly

Table 10.3 Sources and ways of learning.

	Way of learning	
	Excepting rules (single loop)	**Changing rules (double loop)**
Inside the ICJV	Internal autonomous learning	Internal induced learning
Outside the ICJV	External autonomous learning	External induced learning

carry it out. They document the analytical steps better. While in principle the learning stays implicit, it is better accessible for a transformation to explicit knowledge.

 Autonomous learning = (obs.) The planned cycle time for erecting a span of the BangNa expressway was 2 days. There were five underslung girders with crews dedicated to this purpose. All five crews mastered the 2-day cycle in a short time. All crews made improvements but they did not show effect: for this, you have to cut the cycle time to 1 day, as a day is the smallest unit for progress. One of five crews was finally able to achieve it by internal autonomous learning. This is tremendous progress. If the result of the learning, this innovation, would be available for the next project, it would mean that the contractor is able to reduce resources significantly and be more competitive or have larger profits. He could also offer the owner a much shorter construction time. Since the knowledge was implicit and accumulated over time, it was not even possible to transfer it to the other crews without the additional effort to analyze the process and to make it explicit. There was not enough time left in the project to accomplish the knowledge transfer. The successful crew was not able to explain the changes. The crew made small changes throughout $1\frac{1}{2}$ years, slowly, step-by-step with many unsuccessful attempts.

Internal induced learning in the same project took place when developing the design. The owner started an external induced learning process by deciding to use a specific form of financing with promissory notes.

External autonomous learning becomes evident to the ICJV only indirectly, for example, by a changed behavior of the owner during implementation.

My basic motivation for writing this book was the fact that knowledge about construction project management for megaprojects was only available to me implicitly. Your motivation as reader might be that you profit from the explicit discussion. The same situations face contractors who pay attention to learning processes and who want to improve from project to project by learning.

Polanyi (1958) introduced the distinction between implicit and explicit knowledge. Table 10.4 lists the main characteristics of the two knowledge types.

Nonaka and Takeuchi (1995, Figure 10.17) used this dichotomy to model a learning cycle in an organization, in our case, in an ICJV. Let us assume someone gathers knowledge through experience; at this stage, this is implicit knowledge. People with implicit knowledge will watch each other and learn by observation. The information is analog and the learning, autonomous.

By analysis and reflection, a person can structure, generalize, and codify this implicit knowledge and mold it into explicit knowledge. We call this transfer from implicit to explicit, externalization. The information becomes digital.

When two forms of explicit knowledge come together, it is easy for the bearers to communicate. They can combine old and new knowledge to advance understanding.

Table 10.4   Differences between external and internal knowledge.

Explicit knowledge	Implicit knowledge
Inter-subjective, rational, and reflected	Personal, cognitive, and based on experience
Digital information	Analog information
Structured	Unstructured
Fixed content and more generalized	Specific
Easy to codify and document	Difficult to capture and codify
Easy to share, to teach, and to learn	Difficult to share, to teach, and to learn

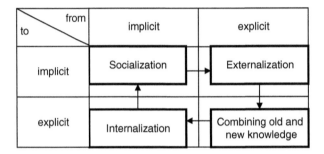

Figure 10.17   Model of knowledge creation.

This is the described communication process between encoder/decoder and the base for induced learning.

When new knowledge becomes routine, then the process of internalization transforms explicit into implicit knowledge. In such a way, a new plateau of knowledge becomes a solid base for further action.

In my descriptive megaproject management model, I define learning as a management task through the development of shared cognitive maps. In this case, induced learning is not an option for managers but a necessary task.

The cycle of knowledge growth follows this path: ICJV members share implicit knowledge on a certain plateau as a management task. Some of the group, especially the project manager, must externalize this knowledge. They compare and discuss the new awareness for creating an innovation. With time and practice, they internalize this knowledge so that it becomes routine; then, they share it through socialization on a higher plateau.

This can work for autonomous as well as induced learning. In the case of autonomous learning, a crew uses some practice. This can be – to continue using the analogy of a bridge – a certain way of erecting the superstructure. They coordinate their work by observation (learning curve). Someone has an idea (why not do this?) and convinces the others to change the established practice. This is externalization. The crew applies the change and it works well (combining old and new knowledge). Then, the changed practice becomes routine. Many such incremental changes can build up to the observed important innovation that reduced cycle time from 2 days to 1 day. The process of externalization through reflection and analysis is not very important for autonomous learning. Only the continued application of the cycle of knowledge growth brings important results. It depends on repetition.

An Asian project manager discussed the advantages and disadvantages of the learning process when adjusting the site conditions.

Learning = (quote) "The Japanese company, they have a big library in their head office; they have manuals for all different problems. When you start the project, you need to do this, do this and do this, how to tackle the problem at the time. When you have very bad concrete surface problem, then you need to do this and you need to do this. They have a very good library. They are depending on their head office very much. In our case we have no written information in the library in the head office, so we have to find a way always newly from the bottom, we have to start building. It has a good side and bad side. We should be, how can I say, working so hard and we should, we could apply the actual site conditions to our own method very well. While they have a very standardized method without considering the real site conditions. So we have hard time, we spend more time to establish the work method."

Wei Han, Taiwanese project manager

Another manager explained that know-how (methodological knowledge) is more important than know-that (factual knowledge).

Methodological knowledge = (quote) "Yeah. I learnt this and finished, another headache, another headache, always like that because I had to handle all the engineering matters. If somebody asked me what is the most valuable knowledge you learn, then I would say: The method of making things a success that is the most important knowledge."

Peng-Chu, Taiwanese project manager

An example of the effects of repetitive learning is the progress curve (Section 13.3). It poses the following tenet: "*When doubling the previously accumulated quantity produced with one technology, it is possible to reduce the costs per unit by 80 to 70 percent in real terms (inflation adjusted).*" This tenet contains a caveat (it is possible), and therefore lacks the character of a natural law, and we should not confuse it with one. "*Produced with one technology*" is one way of describing repetition. There are several reasons prompting the progress curve, one of which is the learning curve:

1. Labor efficiency (learning curve)
2. Standardization and improvement of methods
3. Technology advances

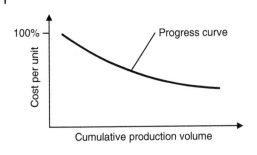

Figure 10.18 Progress curve as manifestation of repetitive learning.

4. Better use of equipment
5. Changes in resource mix
6. Better design

We should not confound the learning and the progress curve. Increases in labor efficiency cause the learning curve. This includes a better organization of the work process and a better crew coordination. Typically, learning curve effects stop after six to ten repetitions; after that, the productivity remains stable.

The progress curve has a mix of causes and productivity will keep increasing. If we forget the caveat (it is possible), we can formulate the tenet as a mathematical function:

$$c(x) = c_1 (1-\alpha)^\delta$$

In this formula, c_1 denotes the costs of the first unit, α the cost reduction factor, δ the times of doubling output, and $c(x)$ the cost of unit x. The exponential function approaches an asymptote (Figure 10.18).

To profit from the progress curve, a contractor must be able to access repetitive projects with repetitive steps.

11

Basic Functions

The basic functions of megaprojects are project knowledge, trust, sensemaking, and commitment. They work like lubricants in an engine. Without a very good dose of these, a megaproject will grind to a halt.

We know the importance of these functions for construction projects. Therefore, we should never forget to manage them actively. Project knowledge is crucial, and it will develop faster when actively managed. This is also true for the other three functions.

Figure 11.1 reminds us of the place of the basic functions in the management model.

11.1 Project Knowledge

Project knowledge is essential. After signature, typically, the bidding team that was in control of the negotiations hands over all explicit information to the project team. The bidding and negotiating process often takes a year, and the owner and contractors build up a large amount of knowledge. Some of this is in writing, some only in the minds of the negotiators (implicit). The parties at each side of the table have a tendency for selective hearing and memory. Thus, the contractual understanding differs. The bidding team has to transfer all this knowledge (explicit and implicit) to the project team. It should be clear that there is not enough time to externalize all implicit knowledge, and, accordingly, there will be a knowledge loss.

Besides this verbal information, the project team will start reading the textual information, i.e. contracts and minutes of meetings. The very first thing that project managers do when they have time is read and understand the contract. While still reading the contract, the project management team must make the first decisions.

The amount of information available is massive and often ambiguous. Compared with the knowledge to finish the project successfully, this massive amount is just the start. Now begins the learning process, and this process generates further information. The more complex the project, the longer will it take to understand it through learning. The project team must thus develop a high degree of cognitive complexity.

Figure 11.2 displays the knowledge gains during implementation. For simplicity, the knowledge is zero at the beginning, neglecting knowledge from the bidding and

Advanced Construction Project Management: The Complexity of Megaprojects,
First Edition. Christian Brockmann.
© 2021 John Wiley & Sons Ltd. Published 2021 by John Wiley & Sons Ltd.

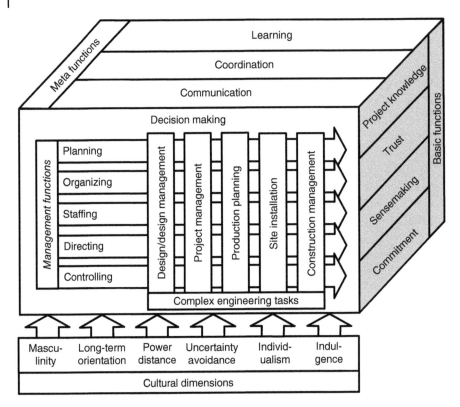

Figure 11.1 Basic functions in the descriptive management model for megaprojects.

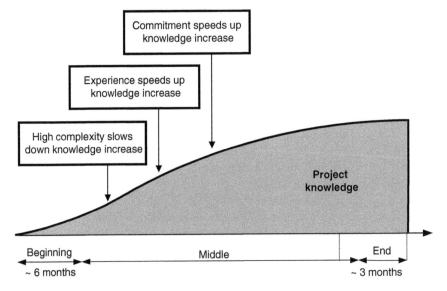

Figure 11.2 Development and influences on project knowledge.

negotiating phase. It reaches 100% at handover; this is also a simplification as learning by the international construction joint venture (ICJV) continues until the end of the maintenance period. The increase in project knowledge starts slowly because much of the information comes in pieces and does not provide a full picture. The more complex the project is, the more diverse and unconnected are the bits of information. This slows down knowledge gains. If the project team fully commits itself to the project, then it can speed up the learning process. Experience helps to connect separate bits of information and speeds up learning as well. However, since experience bridges bits of information based on past know-how, an experienced manager should regard the implicit assumptions he takes as given with suspicion. He must always consider the actual situation of his megaproject.

A megaproject manager explains the process:

Knowledge gains = (quote) "It's impossible. So, the knowledge you have at the end has been mainly determined by the events along the way, events of many kinds. So I think the controlling processes – and it's not only financial controlling, but controlling the whole site what the management does in regard to targets, performances, schedules, schedules for both design and the construction and your whole financial picture, commercial targets and the monitoring. That's where the picture becomes clear. And there is also a point where you are out of your learning curve."

Axel, German project manager

Once a simplified picture emerges, learning speeds up. In the first half of the project, knowledge probably reaches 80%. The project management team does not need much of the missing information until later. It has all the necessary information for the moment. This means that the knowledge gap is the widest during the first half of a megaproject.

A European project manager expresses this in his words:

Knowledge development = (quote) "Yes, yes. So there is, maybe there is even, there you can say, no, just like after fifty percent of the time you should have and you have seventy, eighty percent knowledge of the project. So most of the things need to be clear, outstanding things you have to clarify."

Giorgio, Italian project manager

Unfortunately, we have to make the strategic decisions at the beginning. These are decisions that we cannot correct easily at a later stage. Among them are make-or-buy

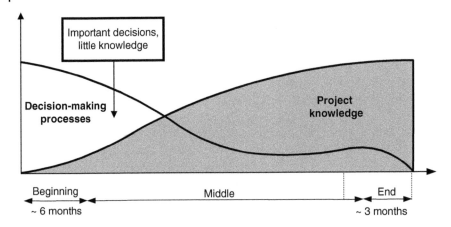

Figure 11.3   Decision-making and project knowledge.

decisions, purchase contracts, construction technology, and site installation. The owner and contractor have already fixed the contract price and time as well as scope by the signature under the contract. Figure 11.3 illustrates the gap between required decisions and knowledge. It is clear that decisions and knowledge have quite different dimensions and that we cannot compare them directly. The figure illustrates a problem but does not quantify it.

The project management team faces the quandary of making decisions with little information on time or making decisions too late with more information. It takes some courage to make the decisions on time, since megaproject managers become more vulnerable to attack. Managers in megaprojects must be willing to make some suboptimal decisions and they need judgment to determine the minimum amount of information required. Intuition must support rationality.

Another European project manager supports this view:

Risky decisions = (quote) "I have seen it with a lot of projects where they made some crucial mistakes in the beginning and to adjust that is very expensive and time consuming."

Andreas, German design manager

11.2   Trust

Trust is essential in megaprojects because delegation is the only way to deal with the very high workload. Delegation without trust means controlling every step of those to whom we delegate work. A project manager who keeps his colleagues in the project management team under strict control is not delegating at all and will become the bottleneck of the project.

Trust = (quote) "Trust in the broadest sense of confidence in one's expectations, is a basic fact of social life. In many situations, of course, a person can choose in certain respects whether or not to bestow trust. But a complete absence would prevent him or her from even getting up in the morning."

Luhmann (2017, p. 5)

The prisoner's dilemma often serves to illustrate the basic function of trust. In this typical example of game theory (Gibbons 1992) two burglars are caught. The police questions them separately, and they can either confess or not. The police require a full confession for the court. The burglars face 1 year in prison if both remain silent and five if both confess. If only one confesses, he will go free and the other, having not confessed, will get a sentence of 20 years. The dominant strategy is confessing for both. Burglar A faces the behavior of Burglar B. If A confesses, he will get 5 years instead of 20 if B also confesses. In case B does not confess, a confession by A will set him free instead of 1 year in prison. Confessing makes burglar A better off regardless of burglar B's decision. The optimal solution for both would be to remain silent, but this requires trust (Table 11.1).

Trusting each other and not confessing would limit the combined sentence to 2 years, i.e. 1 year for each burglar. If both confess, the combined amount would increase to 10 years. If one confesses, the total would reach 20 years. Trust is making life better for teams in difficult situations.

Trust as lubricant = (quote) "Trust is extremely important. If you don't have trust, the joint venture has difficulties to operate because the whole time too much energy and effort is spent on watching what each partner is doing and not devoting it to the outcome of the project."

Walter, German construction manager

Trust and delegation = (quote) "Trust is very important. If I give you authority and responsibility, I should trust you to some extent, otherwise I can't give you anything."

Takeshi, Japanese project manager

Table 11.1 Prisoner's dilemma.

		Burglar B			
		Confess		Not confess	
Burglar A	Confess	A: 5	B: 5	A: 0	B: 20
	Not confess	A: 20	B: 0	A: 1	B: 1

If trust is very important in ICJVs, then the question remains of how it works. Typically, trust develops between individuals through interaction. As the saying goes: *"You have to walk the talk."* In an ICJV, however, there is no time to develop trust gradually; it has to be present at the start. An ICJV has to function on day one; delegation starts on day one.

Trust at the beginning = (quote) "I think what you do, you take trust, you trust everybody at the outset and then you look for the exception. I think that's a very general approach. I think it's a general approach for everybody who worked overseas. People that have not worked overseas would tend to have the reverse, okay? Those that can't trust, quickly get overloaded with work. They end up doing everything themselves."

Roger, American project manager

This quote shows that the relentless pace of megaprojects requires trust from the beginning. There is no time to develop it and no possibility of working without it. This is what Girmscheid and Brockmann (2010) call necessitated trust – trust by necessity and not choice. Necessitated trust is a very special form of trust. Members of ICJVs try to replace it with face-to-face or developed trust based on mutual interaction. The sum of necessitated and face-to-face trust has to surpass a minimum to keep the ICJV functioning (Figure 11.4). A third form of trust is history-based trust, which stems from experience.

Face-to-face trust develops quickly in ICJVs because the members work together closely, often in the same office. However, it is also fragile. When trust is so important in environments with a high workload and the necessity to delegate, everyone will react sharply in case of a breakdown. Sometimes, misunderstandings cause breakdowns and then trust is easily reestablished. However, if misuse of trust occurs, consequences must be strict, or trust will disappear forever.

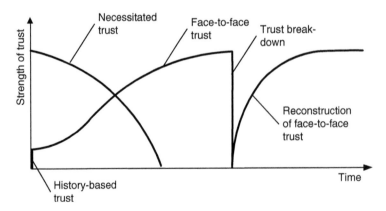

Figure 11.4 Necessitated and face-to-face trust.

Trust development = (quote) "It doesn't develop instantaneously. You know, it's something that, I would say, feeds on itself. You develop a little bit of trust and people start believing in each other and then it really rapidly increases. But it doesn't take much of a hiccup to put you back to square one."

Ernst-Friedrich, German project manager

Figure 11.5 shows a cognitive map of trust in megaprojects. Given a high workload, trust helps increase work output, i.e. productivity.

Perceived competence and relevant experience are helpful for building trust. As trust is two-sided, a positive evaluation of each other becomes important. Opportunism or shirking is not a real option at ICJVs in the beginning when it relies on necessitated trust. It is very hard to hide, as there are not many people in the initial engineering group, and it will very likely lead to a warning, and finally, punishment. Trust together with delegation reduces individual complexity. If a project manager appoints a design manager he trusts, then he can stop worrying about design. Trust helps ferment team-building and reduces controls.

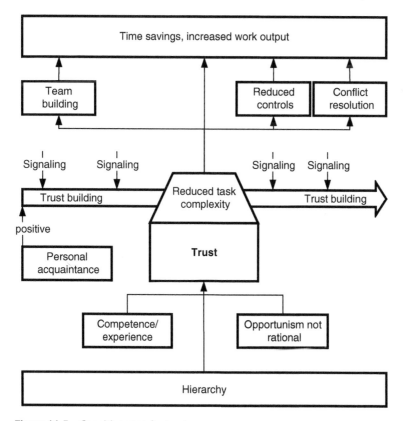

Figure 11.5   Cognitive map for trust.

There will still be conflicts; however, these are easier to resolve in a trusting environment. All this facilitates work and increases productivity.

We need to establish trust by words and actions, and we have to keep the process going by positive signaling (see, you can trust me/yes, I trust you). The required frequency of the signaling is higher at the beginning, but it must continue to the end of the project.

11.3   Sensemaking

Human beings constantly seek understanding and try to make sense. This fact was already one foundation for understanding the cognitive maps of our brains. We actively seek clues to understand what goes on. Human beings cannot endure the absence of sense (Brockmann 2010).

Sensemaking = (quote) "The absence of sense is the horror of the existential nothingness. It is that subjective condition in which reality seems to recede or dissolve completely."

Watzlawick (1967, p. 247)

Sensemaking relies on four assumptions:

1. Human beings search continuously for sense.
2. All individuals are capable of making sense.
3. All individuals have been socialized and have acquired a meaningful set of values and norms.
4. Culture is a coherent system of values and norms.

Thus, sensemaking is context sensitive. What makes sense in one culture does not necessarily make sense in another.

Luhmann (1971) sees sensemaking as the cognitive activity to structure our experience.

Sensemaking in ICJVs = (def.) Sensemaking refers to all interpersonal activities through which members of an ICJV try to structure their human experiences within such an organization. The term "human experiences" includes work and relational aspects.

According to Weick (1995), sensemaking has the following properties:

1. Sensemaking is retrospective since it refers to interpersonal activities that are necessarily passed once we reflect upon them. They must have been encoded previously for us to decode, or they must have been thought out to be encoded.

2. We enact sensemaking, and this means that we create part of our environment as a social construction (Berger and Luckmann 1967).
3. Sensemaking is social action, and as such, it is enacted by groups. This becomes already clear by the use of the sender/receiver model underlying the notions of encoding and decoding. This is the basic model of communication.
4. Sensemaking is continuous, and this finds a parallel in hermeneutics where the model of the hermeneutical spiral is used to illustrate that all interpretations are based on the horizon of previously acquired knowledge. There are no blank sheets. The horizon is the context in which sensemaking takes place.
5. We extract sensemaking from cues. Accordingly, Smircich and Morgan (1982) define leadership as an activity to generate points of reference for sensemaking.
6. Sensemaking is plausible and not exact because the processes of encoding and decoding are not exact, neither is the ensuing whole of what makes sense.

It is clear that communication and action drive sensemaking, and it is a management task to provide clues and direct the shaping of the shared environment (Figure 11.6).

There are four barriers to sensemaking that the management has to overcome: ambiguous or unclear situations, novel situations without structure, lack of group cohesion and acceptance of leaders, as well as conflicts between the sensemaking process and own identity (Büchel et al. 1998). All these barriers are prevalent in the beginning of megaprojects and, during this time, managers must pay much attention to sensemaking. We add decoded sense to our cognitive maps, i.e. we create cognitive maps through sensemaking. The ideal outcome would be overlapping cognitive maps of all ICJV members, reflecting best practice with regard to the megaprojects.

Managers in construction megaprojects will seldom use the word "sensemaking." In the following quote, the reference is to "the same language," which is nothing but the expression of shared cognitive maps.

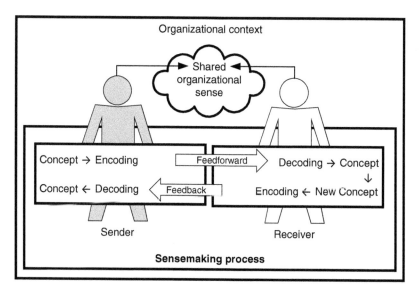

Figure 11.6 Sensemaking through communication.

Sensemaking = (quote) "Yeah, leadership, I think, you know, I think, I can mention this here, you know, when I came here to this project, there were a lot of experts, but there was missing a little bit the combination, the team-work. Everybody was working and, you know, was trying his best, but there was missing the combined achievement of the target. This had to be sorted out very well and also, you know, it is very important, first of all that the joint venture partners are speaking one language."

Hartmut, German construction manager

Sensemaking in construction has one big advantage: the contract defines the scope of the work. There is no search for the main objective, and it is seldom that a member of an ICJV does not accept this objective. However, the goal is far away, maybe 3 or 4 years. Intermediate goals and a path toward the goals need definition and acceptance by sense-making. The most notable barriers in megaprojects are ambiguity and a lack of structure and identification with the project in the beginning.

Successful sensemaking reduces complexity; a majority of individuals share some explanations from the infinite possible ones. The search process ends. The result of successful sensemaking is the development of common goals and processes to achieve these goals. If the members of an ICJV accept the goals and processes, it becomes possible to build a strong identification with the ICJV. Consequently, the team gels together and can provide support for individuals.

Sensemaking as a management process relies on communication and action. Members of the ICJV can observe the actions of the management team and others. The management team needs feedback to understand how the members of the ICJV perceive and evaluate these actions. This happens mostly through personal communication. Then, management must use all possibilities to provide the information that makes the action understandable and allows building shared cognitive maps of the focal megaproject. This is a continuous process requiring unremitting attention. Personal communication, formal and informal meetings, and celebrations and events provide a platform from which we can give the important clues for sensemaking (Figure 11.7).

Difficulty of sensemaking = (quote) "We had many amongst the senior staff, knowing them for years or more being loyal to the company, coming from the home office. But if you have 28 nationalities and two-thirds, I don't know what the numbers are, are hired on the streets, you cannot expect that identity. These peoples just have their jobs."

Thomas, German project manager

Sensemaking in megaprojects is a massive management task.

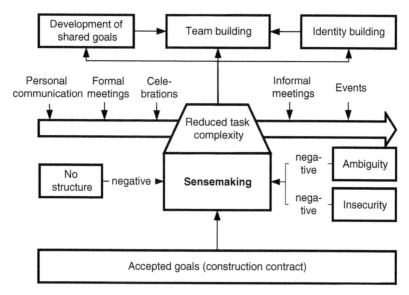

Figure 11.7 Cognitive map of sensemaking.

11.4 Commitment

The members of an ICJV come from different companies and third-country nationals from the international labor market. Many employees from the partner companies have little choice but to accept the secondment, even if they do not like it. Some third-country nationals simply need work. How can we expect to form a high-performance team from such beginnings? How can we assume that the average member of the ICJV works hard with only one goal – the success of the ICJV? How can the top management team generate commitment?

Commitment = (def.) This is the degree to which an employee identifies himself with an organization such as an ICJV. It includes the acceptance of the goals and norms of the ICJV, willingness to engage oneself fully for the ICJV, and strength of the wish to be part of the ICJV.

The development of commitment in an ICJV is a management task; it will not happen by itself. The first step is the agreement between partner companies that the ICJV takes precedence over the short-term goals of the individual companies. Full commitment can contribute to increasing profits for the ICJV. Any deals that partners to the ICJV make for their own benefit will reduce this profit. Worse, a tit-for-tat strategy leads to the looting of the ICJV by all employees for their own company's sake. This leads to a breakdown not only in teamwork but also in trust. It is a vicious circle.

There are no data of the financial effects of looting, as profits in megaprojects depend on many factors, not only on commitment. Anecdotal evidence points to a rather strong tendency toward looting.

 Commitment in megaprojects = (obs.) Two companies from the same country in Africa formed a construction joint venture. The project manager presented an analogy: there is a small chicken and two choices. One can neglect the chicken and pluck every single feather, or one can feed the chicken to turn it into a hen. In the first case, there are some feathers and little meat. In the other case, there are many feathers and a nice roast, minus the expenses for feeding.

The companies liked many feathers and a roast. However, when the first money came from the owner, both bought equipment that would be of good use in the respective companies later. It was only of some use to the ICJV. Individual benefit took priority over shared benefit, and the story about the chicken was just a nice story.

The project manager asked for commitment, and the companies fired him in return.

The principal/agent theory would recommend taking "hostage." The possibility to join forces on future projects could be such a hostage. If one company decides to loot, the other could threaten to forsake future cooperation. Unfortunately, this is not a strong threat as future contracts are not a certainty. Only rationality, common sense, and trust by partner companies can provide the foundations for commitment in ICJVs.

Even if companies agree on commitment instead of looting, it still depends on individuals. Employees seconded to an ICJV will always know that the ICJV will dissolve with handover and that their time in the ICJV is limited. Advancement in ICJVs is also, therefore, limited. In this case, companies can take the career of each employee as hostage by making the success of the ICJV a condition for future advancements in the company. However, goodwill and common sense by every member in an ICJV are the best guarantee.

Bateman and Snell (2002) define high-performance teams as those with a strong cohesion (harmony) and high group norms (targets). From the perspective of the ICJV, harmonious groups lacking ambitions are disastrous: everybody is happy with low output. In this case, it is better to break up the group cohesion and start some disagreement about the norms to follow. Then, there is at least a chance to increase the norms and expectations of oneself and the team. Once a team focuses on high standards, the next step would be creating more cohesion (Figure 11.8). This grid allows analyzing the status of a team and charting the course ahead.

It is clear that management has to promote high professional standards and cohesiveness. In ICJVs, the project manager cannot always determine who joins his team, but he can more or less decide who stays and who he will send back home.

Figure 11.8 High-performance teams.

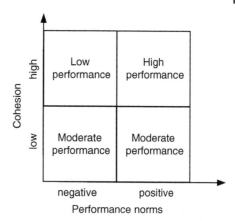

Figure 11.9 Commitment to the parent company and ICJV in case of a hidden agenda.

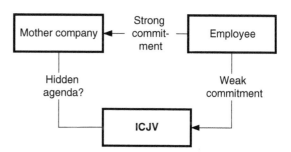

Figure 11.9 illustrates the dilemma of ICJV members. They remain employees of the parent company with their future in that company. If a parent company pursues a hidden agenda (looting), then it also ties the employees to their company by complicity and the commitment to the ICJV will suffer.

Project managers can set an example of commitment and thus lead by example, as the following quote shows.

 Commitment by the project manager = (quote) "Yeah, you will talk to our project manager, if you talk to him you feel quickly what kind of dedication and the character, what personality is driving him. And a lot of these people around who have such a high pride in what they have to deliver that they always push and fight, and never give up, you know."

Stephan, German project manager

It is important to create an identity with the ICJV, as identity strengthens commitment. This is not so difficult since many individuals are proud to take part in a megaproject creating something special.

Identity = (quote) "I think of an identity in the sense that there is a pride and a satisfaction of being associated with a successful joint venture. Especially when they are talking to other people in the industry and find that this project is recognized as being successful, they like to feel to be part of it."

Dave, Australian project manager

Even without success, there remains the pride in the completed megaproject.

 Pride = (quote) "I don't think so much they need the identity of certain organization, I think it's more the identity, in our case it's the identity of having achieved a certain project, a certain, forty kilometers of Taiwan High Speed Rail, we build it. We have one poster, which we did in Athens, where we said: We built the airport."

Andreas, German design manager

Exchange theory can explain commitment to an ICJV and show the road toward it. If an individual feels that he is receiving as much from the ICJV as he gives, then satisfaction will lead to commitment. An ICJV cannot provide for a career directly, but it can offer challenging work, ensure participation in decision-making, provide different tasks on a similar hierarchical level, and generate pride in being part of something outstanding.

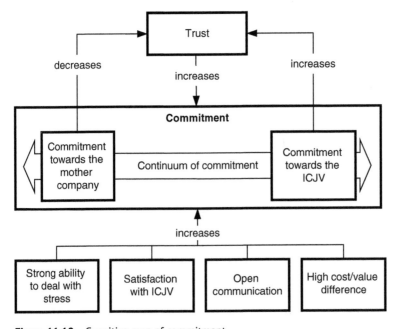

Figure 11.10 Cognitive map of commitment.

Commitment spans a continuum, from very little to a lot of commitment (Figure 11.10). According to the exchange theory, a high cost/value difference furthers commitment, open communication, and general satisfaction or pride in being part of a megaproject. Especially in the beginning of a megaproject, with the lack of structure and the omnipresence of ambiguity, a strong ability to deal with stress will help the commitment process.

Trust between the members of an ICJV increases commitment, while limited commitment undermines trust in turn.

12

Cultural Management

Hofstede and Hofstede (2005, p. 3) provide a misleading definition of culture when they claim it to be the "software of the mind." Software forces all computers to follow exactly the same steps without variation. However, individuals from the same national culture are different. A better construct of national cultures is that of a frame constituted by norms and expectations. Within this frame, everyone is free to place himself or to move about. A small border transgression will entail some nudging back into the frame. Society will punish larger transgressions more severely. We should not think of such frames as rectangular; they can take any shape. Some are small, representing a homogenous culture, and some are large, depicting a diverse culture. Different cultures can overlap to some degree. No culture is superior to another; they are all different. There is a certain set of values that all human beings share; with regard to these, all cultures are the same. Among them is the right to dignity, freedom, to find shelter from the environment, and food. Sometimes, politics pervert cultural values with terrible outcomes; an example is the Third Reich in Germany. There is no shortage of other historical examples. In such cases, the normative behavior of peoples can become unacceptable. Diversity is not an excuse for wrongdoings.

It is a misconception that members of ICJVs see collaboration between different national cultures as a big problem. The opposite is quite true; cultural diversity helps find solutions that are more adequate. Nevertheless, few would deny the absence of negative side effects due to cultural misunderstandings.

Positive cultural influence = (quote) "It has a positive impact if they work together. I mean, if you look here, especially if you look at the project, which we did here; there we had different cultures, we have Japanese, we have Germans, we have Thais, yes, and I think they work all good together. In my opinion, otherwise the result would be not as it is now."
Ernst-Friedrich, German project manager

When working in an intercultural environment, learning and the ability to adjust are very important. When working outside the home country, the old adage still holds true: *"When in Rome, do as the Romans do!"*

Advanced Construction Project Management: The Complexity of Megaprojects,
First Edition. Christian Brockmann.
© 2021 John Wiley & Sons Ltd. Published 2021 by John Wiley & Sons Ltd.

Cultural adjustment = (quote) "For my case, I try to adjust myself to the new circumstances, otherwise I cannot survive in a different country. But some people are really stubborn."
Son, South Korean project manager

Hofstede and Hofstede (2005) assume the following axioms when discussing culture:

- Culture is learned.
- Culture is a social endowment.
- Culture is relatively stable in time.
- There are national, branch, company cultures, and other cultures.
- Culture describes groups, not individuals.
- Culture determines the perception of each individual.
- Culture manifests itself in symbols, heroes, rituals, and values.

We learn culture through primary (family) and secondary (society) education. The training starts the first day of our life and continues to the last. People might eat Big Whoppers instead of local food. However, this is not a change in underlying cultural values. Pride in one's national cuisine will stay untouched by Big Whoppers, and mama is still the best cook in the world. Culture is not limited to national cultures. Besides branch and company cultures, there are plenty of subcultures with their own sets of values, such as punk culture or the culture of the voluntarily homeless.

I have already introduced and discussed Hofstede's six cultural dimensions, i.e. power distance (PDI), uncertainty avoidance (UAI), individualism (IDV), masculinity (MAS), long-term orientation (LTO), and indulgence (IVR). All of these influence the descriptive megaproject management model (Figure 12.1).

As stated before, cultures overlap to some degree. If we take Hofstede's dimensions and map the values for each of the three different national cultures (Germany, Japan, and Thailand), we can see that the commonalities between the two Asian countries are not necessarily greater than between a European and an Asian country. There is no cultural East/West divide (Figure 12.2).

A similar picture emerges when comparing values from more countries (Table 12.1; Hofstede and Hofstede 2005). A clear separation between East and West emerges only for LTO. PDI is high in Asia as well as in France. IDV is high not only in the West but also in Japan. The MAS is almost the same in France and Taiwan. UAI has the same values in France and Japan as well as in Germany and Thailand. Finally, IVR is low in Mainland China and considerably higher in Taiwan.

We should not interpret the values as exact. The research methodology and the concept of measuring culture with six dimensions do not warrant this. They rather have an indicative value. I would consider data that vary by not more than five points as similar. Only greater differences deserve attention.

McSweeney (2002) calls Hofstede's model a triumph of faith and a failure of analysis; this is a scathing critique and merits observation. However, national culture is a construct that is an obvious interpretation of phenomena observable in ICJVs; people see differences and

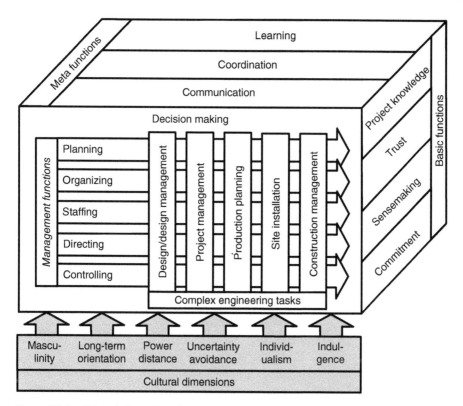

Figure 12.1 Cultural dimensions in the descriptive management model for megaprojects.

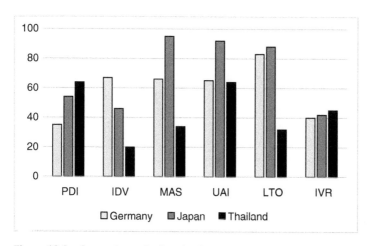

Figure 12.2 Comparison of cultural values.

Table 12.1 Values for Hofstede's dimensions.

	PDI	IDV	MAS	UAI	LTO	IVR
China	80	20	66	30	87	24
France	68	71	43	86	63	48
Germany	35	67	66	65	83	40
Japan	54	46	95	92	88	42
Netherlands	38	80	14	53	67	68
Switzerland	34	68	70	58	74	66
South Korea	60	18	39	85	100	29
Taiwan	58	17	45	69	93	49
Thailand	64	20	34	64	32	45
UK	35	89	66	35	51	69
USA	40	91	62	46	26	68

can only make sense of these differences by referring to the construct of national culture. They understand it as a given. Differentiating between an individual and a group from one national culture and accepting a fuzziness (frame) helps manage differences.

The advantage of Hofstede's model is that it refers to the belief of members in ICJVs in the construct of national cultures and provides points for discussion. The dimensions are relevant for managing in an international environment. The model allows for a practical approach to cultural differences by initiating discussion on management options.

I think we can remove cultural barriers or what we perceive as such only through communication based on an initial analysis of perceived cultural differences. The model by Hofstede and Hofstede provides such differences. We can perform a cultural analysis by establishing a framework in the form of a matrix of the cultural dimensions and management functions, meta-functions, basic functions, or components of project management as defined in the PMBOK, such as risk, human resources, or scope.

Figure 12.3 shows a matrix of cultural dimensions and management functions. The question mark in the cells asks for the influence of cultural differences on these functions. The dimensions of national culture are an independent variable, and management functions are dependent variables. Instead of management functions, we can choose any dimension meriting cultural analysis.

In a first step, for each cultural dimension, we can introduce the values from Hofstede for the national cultures represented in the top management of an ICJV. This will most often be cultures from countries wherefrom the partners originate. Figure 12.4 depicts the values for two countries, i.e. Thailand and Germany. Here, a first discussion between the partners can ensue: Do the partners feel that the values from Hofstede represent their attitudes and behavior? Are the dimensions relevant? Do Hofstede's dimensions represent all relevant aspects? If the partners answer some of these questions with a "no," then they are free to change the values, drop, or add dimensions. More important than the values is the process of the discussion; it is a process of sensemaking and of creating a shared understanding.

	Planning	Organi-zing	Staffing	Directing	Control-ling
PDI	?	?	?	?	?
MAS	?	?	?	?	?
IDV	?	?	?	?	?
UAI	?	?	?	?	?
LTO	?	?	?	?	?
IVR	?	?	?	?	?

Figure 12.3 Influence of cultural dimensions on management functions.

		Planning	Organi-zing	Staffing	Directing	Control-ling
PDI	T: 64 G: 35		+++ T:↓		+++ T:↓	+++ T:↓
IDV	T: 20 G: 67		++ G:↓		++ G:↓	++ G:↓
MAS	T: 34 G: 66			++ G:↓	++ G:↓	
UAI	T: 64 G. 65	+++ ☺		++ ☺		+ ☺
LTO	T: 32 G: 83	++ G:↑	+ G:↑			

Figure 12.4 Cultural analysis of a Thai/German ICJV.

In sum, this is the discussion of the dimensions as independent variables and their respective values.

The second step is an analysis of the impact of a difference in one cultural dimension on one management function. It concerns all cells of the matrix and determines the dependent variables in the cell. There are three possibilities: when there are no cultural differences, then there is no impact; if there are differences, there might or might not be a causal link; if there is a causal link, then we must determine the strength.

According to my own analysis, there are the following impacts:

- Planning depends on UAI and LTO. High UAI leads to a preference for formalized planning as planning anticipates the future. LTO influences how we implement plans. A high LTO leads to a pragmatic approach; if the plans are deficient, managers with a high LTO will abandon them. It is important to reach the goal. The reaction to changes is quick, and managers do not understand planning assumptions as fixed; they will control them.

- Planning is contingent on PDI, IDV, and LTO. PDI determines the number of hierarchical levels; IDV influences the preferred organization of work in teams or by individuals; LTO affects the arrangement of the workplace, e.g. Asian managers often observe feng shui principles.
- Staffing deals with individuals, and national cultures do not describe individuals. However, we can use the cultural dimensions to describe desired attitudes for members of ICJVs. In megaprojects, the choice of the right people demands the ability to endure stress, maintain personal balance, intercultural competence, ability to communicate, empathy, and ambiguity tolerance. All these characteristics can in part find explanation in the cultural dimensions. IVR helps deal with stress; femininity strengthens personal balance, intercultural competence, communication patterns, and empathy. A low UAI makes it easier to deal with ambiguity. The characteristics that Hofstede describes by the somewhat misleading descriptive term "femininity" are especially helpful.
- PDI, MAS, and IDV influence directing. The value of PDI determines the preferred leadership style on the continuum from authoritarian to consultative. MAS predisposes the comportment on a continuum from caring to demanding behavior, and IDV regulates the amount of teamwork.
- PDI, IDV, and UAI govern controlling. A high power distance allows for strict supervision; a high collectivism also accepts strict controls, as everybody knows his place in a hierarchy; a strong uncertainty avoidance leads to a different view of controls. On the one hand, the superior shows an interest in the work of the subordinate, and, on the other, it allows for shifting responsibility. As soon as a superior understands a problem, he also owns it.
- Indulgence does not have a direct influence; it works like a moderator variable.

The result of this subjective analysis indicates that we have to take care of the following cells:

- PDI: Organization, directing, and controlling
- UAI: Planning, staffing, and controlling
- MAS: Staffing and directing
- IDV: Organization, directing, and controlling
- LTO: Planning and organization

These identified cells show some content in Figure 12.4. The third step of the process is to determine the strength of the impact, and this is indicated in Figure 12.4, on a scale from weak (+) to strong (+++). Again, this is a subjective assessment.

The fourth step combines cultural differences and strength of impact. In case of large differences and strong impacts, we have identified the most critical cells; they are highlighted by a thunderbolt because they might indicate a disruptive conflict. When there are no cultural differences, a smiley designates "no problem."

The fifth step is the most important, and it contains two levels of discussion. First, the subjective analysis requires discussion and, most likely, changes. If the top management agrees on an outcome, the analysis becomes intersubjective. Second, the top management must agree on the action for each problem identified in a cell. In the given example, the Thais might choose against their cultural attitude to work with lean management and establish flat hierarchies. Consequentially, the organization reduces the levels of hierarchies, the management style becomes more consultative and controlling more empathetic. As Germans use such an approach more often, they can help their Thai colleagues to adjust.

On the other hand, Thais can help Germans to choose employees with a feminine streak and create a more empathetic leadership (people orientation). Thais can also help Germans to profit better from a collective work approach, strengthening teamwork. Finally, they can help Germans to take a more pragmatic approach to planning and to pursue long-term goals instead of short-term goals.

I repeat, whatever the outcome of the discussion, the proposed cultural management approach relies on open communication. Members of the top management from different cultures will profit from the discussion and identify problems and solutions.

Advantages of cultural diversity = (obs.) Plans are anticipating the future, but they never are the future. Thus, all plans become problematic at some points in the future. Western managers tend to stick with the plan and try to remove every obstacle to the plan. Asians tend to walk around the problem and leave it unsolved, concentrating on the ultimate goal. Asians sometimes laugh at Westerners when they work hard to remove an obstacle. Westerners sometimes complain that Asians run away from the smallest problem by sidestepping it. Experienced intercultural managers see an opportunity to learn. If the problem is small, it is beneficial to stick to the plan. If it requires a lot of effort to solve it, it would be better to sidestep. One cultural approach is not superior to another. Knowing different cultural approaches enables us to solve problems more adequately.

The following observation demonstrates how cultural differences in power distance can affect the organizational structure.

Power distance and organization = (obs.) A Western project manager proposed introducing lean hierarchies on the secretarial level for a project in Asia. He suggested a pool of three secretaries for the four top management positions. The argument was that they could help each other whenever one manager had a high workload for them. The Asian partners agreed. After 3 months, however, it became clear that the two Asian managers had talked to two of the three secretaries and asked them to become their personal secretaries. It was not acceptable to these managers not to have "my secretary." The pool shrunk to one secretary. Luckily for the two Western managers, she was the most efficient one.

The Western project manager made the mistake of not understanding the influence of power distance on this organizational problem and not discussing the problem openly beforehand.

13

Innovation in Construction Megaprojects

Many academics criticize the construction industry for its purported lack of innovation. Many practitioners agree with this interpretation. Yet, this industry designs and builds the largest projects of the world – construction megaprojects. This does not make sense. Instead, the discrepancy begs the question whether the lack of innovation is a valid description of the construction industry. A megaproject seems ideal to test the construction industry for its innovation potential. The collection and analysis of data through longitudinal action research combined with a case study approach provides the required level of detail and an understanding of interdependencies. The BangNa Expressway in Thailand serves as the focal case; it is the largest bridge in the world. Brockmann et al. (2016) have previously published the contents of the following Section 13.1 as a journal article.

The descriptive megaproject management model is the overall cognitive map of a megaproject and serves as the orientation for a given project. Much of innovation happens between two projects and cannot be depicted within the model. However, learning is part of the model and closely related to innovation. Figure 13.1 shows how innovation transforms our cognitive map from one project to the next. The figure shows a very simplified idea, where the transformation affects the complete model. In reality, it might just influence parts of the model.

I have said before that learning accompanies the innovation journey. For my purposes, learning describes the process, and innovation describes the result. An invention is an intermediate result without a business use.

 Invention in construction = (def.) Creation of new ideas or things; in the business environment, this means changes to products and processes from within a company or joint venture. These changes have different levels of impact, and they can be new to a company, the construction industry, or the world, and they may take the form of an incremental, modular, architectural, system, or radical change. An invention includes acceptance of the new product or process by the company or joint venture; it excludes adoption by the market.

Advanced Construction Project Management: The Complexity of Megaprojects,
First Edition. Christian Brockmann.
© 2021 John Wiley & Sons Ltd. Published 2021 by John Wiley & Sons Ltd.

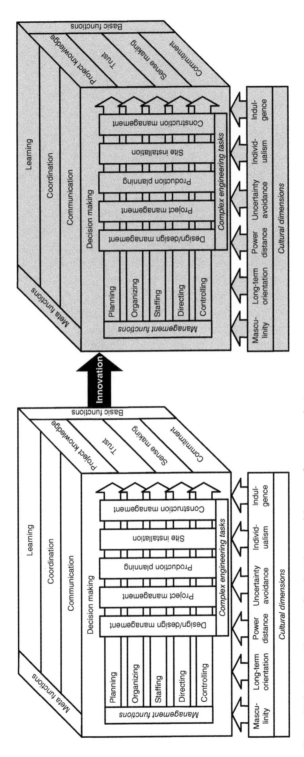

Figure 13.1 Innovation transforming the cognitive map for megaprojects.

13.1   Aspects of Innovation

Some of the academic authors who strongly deny the innovativeness of the construction industry are Laborde and Sanvido (1994), Koskela and Vrijhoef (2001), Woodhuysen and Abbey (2004), and Reichstein et al. (2005). Another group, prima facie a minority of researchers, have come to the opposite result (Dibner and Lemer 1992; Slaughter 1993; Nam and Tatum 1997; Slaughter and Shimizu 2000).

Considering iconic buildings and structures such as the tallest high-rise in the world (Burj Khalifa), the longest spanning bridge (Akashi Kaikyo Bridge), the longest rail tunnel (Gotthard Base Tunnel), or the tallest dam (Jinping-I Dam), it is difficult to imagine that these projects are not highly innovative. Many historic structures serve as points of identification for metropolitan communities, such as the Golden Gate Bridge for San Francisco or the Eiffel Tower for Paris. Is it not true that they capture our imagination because they are inherently innovative? Innovation in construction from this perspective seems highly probable.

From a different perspective, innovation looks highly improbable: What innovation can we expect when building a garden wall, a garage, a one-family home, the ubiquitous small office, or an industrial building? Renovation is another domain of the industry that, by definition, requires low levels of innovation (Winch 2003) while in many countries, the larger part of construction investment is spent on it (Vainio 2011). Constructed products vary vastly from the replacement of a door handle to a multi-billion-dollar oil refinery. Perhaps we cannot make a statement about the innovativeness of the construction industry as a whole, and instead must distinguish between degrees of innovativeness for different types of constructed products.

Research in construction innovation started more or less 30 years ago with a series of contributions from Tatum (1986 onwards). While research was concentrated in the beginning on products and construction technologies, attention has since shifted to processes and organizational change (Gann 2003). However, even today, few case studies have been published pinpointing and naming construction innovations at the project level (Johnson and Tatum 1993; Slaughter 1993; Slaughter and Shimizu 2000; Gambatese and Hallowell 2011; Ozorhon 2013). None of these describe project innovations in detail, nor do they follow the relationships among them closely. In sum, we lack knowledge on the details of innovation in construction projects.

This requires answers to the following questions: (i) What kind of and how many innovations can we identify, and how are they generated? (ii) How are these innovations interrelated? (iii) Are the innovations new to the company, the industry, or to the world? (iv) Who initiated the innovations?

Economists define *technology* as everything that improves output while keeping inputs constant (Taylor 1995). In the simplest form of a production function (Figure 13.2, cf. Section 8.3), the output or yield (Y) depends on the input of two resources only – labor (L) and capital (K) – while technology (A) acts as a moderator variable: $Y = A \times f(L, K)$. It is easily conceivable to expand this production function by other inputs such as human or social capital if desired.

The global technology space is the n-dimensional space of all available technology. Innovation allows us to expand this technology space and shifts the technology frontier while

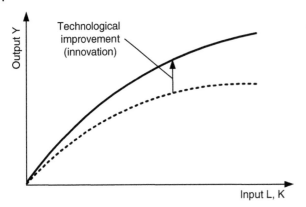

Figure 13.2 Increased output due to innovation.

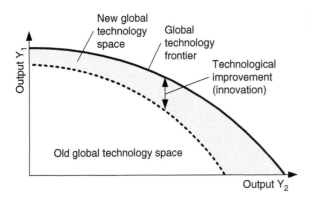

Figure 13.3 Technology space and technology frontier.

keeping the inputs (e.g. L, K) constant. Efficient production is only possible on the technology frontier at the global level. A production level inside the technology space is inefficient because it uses the same amount of inputs to produce a smaller output. Figure 13.3 shows the case of two outputs. One (Y_1) could be the total square meters of building output, while the other (Y_2) could represent a bundle of all other outputs such as food, cars, and clothes.

The discussion of efficiency depends on the two assumptions of perfect information and global access to the full technology space. These assumptions seem very idealistic in the regional construction markets, which we encounter most of the time. On regional markets, an innovation for one firm can enlarge the locally available technology space. Some innovations affect only the construction industry and shift the frontier within the global technology space by providing an improved input/output relationship in construction alone. Given this discussion, it seems helpful to draw a distinction between global, industry, and company innovations. A company is always the innovator in this scheme, and if the innovation is global, it is new to the industry as well.

Slaughter (1998) has introduced five types of innovation based on the work of Henderson and Clark (1990) that have varying impact: (i) incremental change, (ii) modular change, (iii) architectural change, (iv) system change, and (v) radical change. Incremental changes have the smallest impact, which are inherent in all design and construction processes and can stem from basic research. This could be a reduction in rebar weight relative to concrete volume ($kg\,m^{-3}$) due to more appropriate computational assumptions. Modular changes exert a broader influence but are still limited in their impact. An improved formwork system to erect bridge columns may serve as an example. Architectural changes affect other parts of the structure because of existing interrelations between components. Bridge bearings transfer forces from the superstructure to the columns. Any improvements in the design of the bearings will have effects on both the superstructure and the entire substructure (not only on the columns). System changes impact the overall system. All construction methods belong to this group, as the design of the structure, its costs, quality, time, resources, site installation, make-or-buy decisions, and project organization are affected. Radical changes occur seldom and transform the overall approach to particular problems. Some new construction technologies are not only systemic but also radical: the segmental bridge construction technique was such a radical innovation several decades ago.

Nam and Tatum (1997) promote the idea of innovation champions, and they distinguish between technical, business, and executive champions. Within their sample of champions from 10 innovative projects, only one is a project manager; the others are all vice presidents or higher up in the hierarchy. In a megaproject, different groups are involved that might promote innovation: (i) top management at home (in the case study, this is the director of the international division), (ii) leader of the bid and negotiation team, (iii) project management, (iv) construction management, (v) project engineers, and (vi) work crews.

Freeman and Soete (1997) look at the outcome when defining innovation as the actual use of a nontrivial change or improvement in a process, product, or system that is novel to the organization developing the change. The relevant system in construction is the project, and we can distinguish between innovations within the technical, management, and legal organizations of the project. Summarizing the ideas presented so far, we can formulate the following nominal definition for our purposes to guide data evaluation.

 Innovation in construction = (def.) Changes leading to an improved input/ output relationship for products and processes as well as changes within the technical, management, or legal organization of a project that can be evaluated monetarily. These changes have different levels of impact, and they can be new to a company, the construction industry, or the world, and they may take the form of an incremental, modular, architectural, system, or radical change. Champions can belong to the top management, the bid team, or the project (management, construction, engineers, or work crews).

This is one of the many definitions of "innovation"; Baregheh et al. (2009) provide an overview of many different viewpoints. For example, not all include the demand for a monetary evaluation. However, I feel that the above nominal definition serves the purposes of the construction industry best.

13.1.1 Methodology and Case Study Choice

With regard to methodology, Tushman and Nelson (1990) argue that technological change occurs over time and accordingly call for longitudinal studies. Winch (1998) stresses that we need case studies describing innovations and their trajectories. Very few long-term studies following trajectories are available; these include a study by Lu and Sexton (2006) of an architectural firm and two by Harty (2005, 2008) on the implementation of Building Information Modeling at Heathrow's Terminal 5.

It seems we need to go back to grassroots research in long-term case studies to identify discrete innovations and their trajectories. This has become even more urgent since Abbott et al. (2007) described "hidden innovations." They identify four types: (i) innovations excluded from traditional measurement, (ii) innovations without a major scientific or technological basis, (iii) innovations from novel combinations, and (iv) incrementally every day locally developed small local-scale innovations. Cross-sectional studies with the help of surveys or interviews have proven to be inadequate to detect these hidden innovations.

A theory, according to Popper (1999), is a net that we cast to capture the world. When we cast a net with a five-centimeter mesh, we will not catch any fish smaller than five centimeters. We can also use the idea to describe methodology. When evaluating the status of published information on construction innovation, it seems that we might have been fishing with too wide a mesh.

Surveys and interviews on innovation in construction are first problematic with regard to language. Innovation comes in very different styles. Baregheh et al. (2009) collected some 60 definitions from a literature review not limited to construction innovation. Readers who peruse the book on innovation by Loosemore (2014) continuously face the lingering question of what he considers an innovation in each specific context. If this is not clear to the academic, how can it be clear to the practitioner?

The definition by Freeman and Soete (1997) does not give clear guidance to identifying it. They declare an innovation to be a nontrivial change and improvement in a process, product, or system that is novel to the institution. Without a definition of "nontrivial," "change," and "improvement," let alone "system," this demarcation is not operational. The same holds true for the definition of Van de Ven (1986) as "management of new ideas into good currency." Here, we can look at a multitude of definitions for "management" and "idea" and a wide range of interpretations of "good currency." OECD (2005) provides a third definition: "*An innovation is the implementation of a new or significantly improved product (good or service), or process, a new marketing method, or a new organizational method in business practices, workplace organization or external relations.*" It would be interesting to listen to a group of academics determining "significantly" or "business practice." For clarity, I have therefore introduced a definition in the preceding text.

Conducting surveys and structured or semi-structured interviews means – in descending order – to fish with a wide mesh. In studies that take place at a specific time of a project, the researcher requests the interviewees to remember all innovations that have occurred in a project, or worse, in an organization. This will reveal nothing more than what psychologists call top of the mind associations.

If the objective is to discover innovation in construction projects, then long-term case study research through participant observation or action research is more promising. Participant observation leaves the researcher outside the action, and the researcher is at risk to miss important information. Action research is research in action or participative and concurrent with action (Lewin 1946; Coughlan and Coghlan 2002). It provides the deepest possible insight, albeit with the risk of bias (getting too much into the action). However, bias is hardly as much a problem as a lack of information when identifying innovations and following trajectories. A long-term case study with action research seems to provide the finest mesh we can think of.

Important barriers to innovation given in the literature are project size, separation of design and construction, fragmented supply-chain, technical competence of the general contractor, a conservative owner, as well as regulations and standards. How do such barriers affect megaprojects?

Project and company size both influence the amount of slack in an organization (Sexton and Barrett 2003; Gambatese and Hallowell 2011). Megaprojects allow higher investments (fixed costs) if this leads to reduced variable costs. The reason is that the break-even point between (1) low investments and high variable costs and (2) high investments and low variable costs depends on volume (the number of units with variable costs). Specific investments are required for some innovations, and these are often economic for megaprojects.

The separation of design and construction puts an innovation barrier between the two phases (Blayse and Manley 2004). The fragmented supply chain has a tendency to confine innovation (Walker et al. 2003; Harty 2005). A lack of technical competence on behalf of a management contractor who therefore depends on subcontractors for this prevents the integration of innovative approaches and strengthens the departmentalization of innovation (Tatum 1987; Bröchner 2010; Winch 1998). A design–build contract with few subcontracts and a technically strong contractor helps avoid such problems.

Often, researchers present the influence of the owner as crucial (Nam and Tatum 1997; Manley 2006). Some authors consider regulations and standards as boundaries to innovation, mostly hampering it (Gann and Salter 2000; Dubois and Gadde 2002). Both factors shape the external environment for project innovation. An innovative owner will foster and a conservative one will limit innovation as long as he wields a decisive influence. Hence, an owner that does not exert strong influence provides the designer and contractor with the freedom to innovate. We can always circumvent standards by pursuing approval for the individual case. This is, however, only economical for large-scale projects.

The project of choice is the BangNa Expressway in Bangkok, Thailand (Figure 8.26). At the time of completion in 2000, it was the longest bridge in the world (Brockmann and Rogenhofer 2000). The expressway is 55 km long, more than 27 m wide and 18 m above grade as a minimum, all of which translates into a deck area of 1 900 000 m^2, a concrete volume of 1 800 000 m^3, rebar steel in the magnitude of 180 000 tons, and post-tensioning steel with 50 000 tons. It was a megaproject which allowed for specific investment to innovate and for requesting specific approvals deviating from prevailing regulations and standards. The contract was a FIDIC design/build contract based on a conceptual design with few constraints. The owner refrained from exerting a strong influence throughout the design process so that the design/build contractor had every freedom to develop an innovative design. The main contractor was Bilfinger Berger from Germany, at the time, the eighth largest international contractor in the world with substantial technical competence, partnering with CH. Karnchang as the local contractor. None of the civil works was subcontracted; therefore, the supply chain was not fragmented, and the joint venture apparently deemed to possess the technical competence required for innovation. The contractors considered segmental bridge construction (Podolny and Muller 1982) most beneficial. Design knowledge came from one of the most experienced firms in segmental bridge construction,

Jean Muller International, USA. While there is never an ideal case, the BangNa Expressway comes close.

The period for the action research lasted from the beginning of the project in March 1995 until June 1998. Follow-up interviews went on till 2019 to allow verifying the performance of the bridge over time (there were no problems with the quality recorded until 2020).

13.1.2 Innovations and Trajectories

Having a complete discussion of the innovative features of a project means detailing the trajectories of the innovations, i.e. the genesis of each innovation. This is seldom possible because the genesis of many innovations remains in the dark. The BangNa Expressway is fortunately a well-documented project, and much information is publicly available, albeit some of it is not in English.

There are innovations to product and process as well as to the technical, management, or legal organization of the project. I have numbered the innovations consecutively for ease of identification. Establishing and then reading the list of innovations in Tables 13.1–13.5 gives the impression that the whole project is one integrated innovation, and this is a very valid perspective. The chosen framework – and all other imaginable frameworks – is to a large degree artificial and does not capture the essence of innovation in the project. For descriptive purposes, this is helpful; for understanding, it better to view the project as an integrated innovative solution.

Table 13.1 Product innovations, BangNa Expressway.

	Description	Level	Type	Source
1	Mainline columns (Brockmann et al. 2000)	w	S	PM
2	Inclined elastomeric bearings (Krill et al. 1999)	w	S	PM
3	Post-tensioned columns (Brockmann et al. 2000)	c	M	PM
4	Piers without crossbeam (Brockmann et al. 2000)	w	A	PM
5	Web inclination for D6 segments (Brockmann 2000)	w	S	TM-pre
6	Shear transfer (Brockmann 2000)	w	S	TM-post
7	Transversal post-tensioning layout (Brockmann 2000)	w	A	PM
8	D6 segments (Brockmann 2000)	w	S	TM-pre
9	Longitudinal post-tensioning layout (Brockmann 2000)	w	A	PM
10	Bowing effect during match-casting (Brockmann 2001)	w	S	TM-post
11	Mainline portal columns (Brockmann et al. 2000)	w	A	PM
12	Precast portal beams (Brockmann et al. 2000)	c	A	Eng
13	Welded spun piles (Brockmann et al. 2000)	c	M	CM
14	Reduction of rebars for foundation	c	I	PM
15	Traffic control system	c	M	PM
16	Toll collection system	c	M	PM

Table 13.2 Construction technology innovations, BangNa Expressway.

	Description	Level	Type	Source
17	D6 underslung girder (Prade et al. 1998)	w	S	TM-pre
18	Swivel cranes (Prade et al. 1998)	c	M	Eng
19	Chassis for segment transportation (Prade et al. 1998)	c	M	Eng
20	Formwork for piers	w	A	CM
21	Molds for D6 segments (Brockmann et al. 2000)	w	S	CM
22	Computer software for surveying of D6 molds	w	A	CM
23	Construction speed (Brockmann et al. 2000)	w	S	TM-pre, Crew
24	Transport trailers (Brockmann et al. 2000)	i	A	CM
25	Shuttle lifts in precast yard (Brockmann et al. 2000)	i	A	CM
26	Double stacking of segments in storage	c	M	PM
27	Production line in precast yard for D6 segments	c	A	CM
28	Yard as consignment center	c	S	PM
29	Production line in precast yard for portal segments	c	S	Eng
30	Portal beam erection girder (Prade et al. 1998)	w	A	Eng
31	Lifting devices for D6 girders (Prade et al. 1998)	w	M	Eng
32	Quality and quantity of own equipment	c	A	CM
33	Pile-driving equipment	C	A	CM

Table 13.3 Innovations in the technical organization, BangNa Expressway.

	Description	Level	Type	Source
34	Largest bridge in the world (Brockmann et al. 2000)	w	S	Bid
35	Design/build megaproject (Brockmann et al. 2000)	c	S	Bid
36	Production flow in precast yard (Brockmann et al. 2000)	i	S	CM
37	Largest precast yard in the world (Brockmann et al. 2000)	w	S	PM
38	Limitational production function	c	S	TM-pre
39	Simultaneous production with 11 girders	w	S	PM
40	Self-sufficient with regard to rebars and concrete	c	M	PM
41	Full-scale test-span (Fischer et al. 1998)	w	S	PM

13.1.2.1 Product Innovations

I have identified 16 major product innovations, and most of them are interdependent (Table 13.1). Abbreviations used for the three levels of innovation are as follows: world (w), industry (i), and contractor (c). For the five types of innovation, these are incremental (I), modular (M), architectural (A), system (S), and radical (R), and for the source of innovation, top management before signing of contract (TM-pre) and after signing the contract

Table 13.4 Innovations in the management organization, BangNa Expressway.

	Description	Level	Type	Source
42	Density of works	w	S	PM
43	Amount of resources	c	S	PM
44	All civil works done without subcontractors	c	S	CM
45	Joint management for all works	c	S	TM-post
46	Decentralization of management	c	S	PM
47	Matrix organization for construction works	c	A	PM
48	Design coordination for offices on three continents	c	A	PM
49	Start-up company for spun pile production	i	M	TM-post
50	Start-up company for post-tensioning production	i	M	TM-post
51	Commissioning, handover, and opening in eight phases	c	S	Bid
52	Coordination with simultaneous works by DoH	i	S	PM

Table 13.5 Innovations in the legal organization, BangNa Expressway.

	Description	Level	Type	Source
53	Design/build/finance contract	c	S	Bid
54	Independent financing contract	c	S	Bid
55	Payment in promissory notes	i	S	Bid
56	Duration of joint venture	c	A	TM-post
57	Claim over 280 million USD	c	A	PM
58	Litigation in Supreme Court of Thailand	i	A	TM-post

(TM-post), the bid team (Bid), project management (PM), construction management (CM), project engineers (Eng), and work crews (Crew).

The contractor collaborated with the designer to develop the overall design for the design/build contract during the tender stage. The goal was to minimize cost while allowing for completion within time (42 months), at the same time ensuring longevity. In the beginning, the contractor studied two options – segmental bridge construction vs. prestressed longitudinal girders with concrete decks cast in-situ in depth. He made the final decision in favor of segmental construction with external post-tensioning. This required a precast yard and match-cast technology, where segments are cast against each other to guarantee perfect fit. With the precast yard, it became possible to produce more than 50% of the concrete quantities off-site. The large quantities required setting up the world's largest precast yard. Off-site production allowed for parallel production, thus speeding up construction and training of unskilled workers as the tasks became highly repetitive. The numbers in brackets in the following text give cross-references to the listing of the innovations in Tables 13.1–13.5.

Road transportation limited the size of the segments (width of 2.55 m) and determined the bearing capacity of all handling equipment. The width of the expressway (27.20 m) was a demand in the conceptual design by the owner as well as a limitation to the width of the foundation (5 m) and the general height of the bridge (18.60 m). The designer had to develop the cross-section from this small foundation. In addition, he had to consider an opening at the bottom of the piers as they stood in the dewatering ditch of the highway below the expressway (Highway 34). The maximum possible spreading of the columns with regard to rebar placement from its base of 4 m was 7 m at the top [1]. This decision necessitated inclined elastomeric bearings [2] to further limit rebar content by introducing only normal forces in the upper arms of the columns for dead load in addition to the use of post-tensioning for the columns [3]. Standards or experience with inclined elastomeric bearings when designing the piers did not then exist anywhere in the world (Brockmann and Rogenhofer 2000). At the same time, crossbeams at the top of the columns were highly undesirable because they would have blocked the launching of the underslung girders [4].

The top of the columns with a width of 7 m equaled the bottom slab of the D6 segments. Given a reasonable height (2.60 m), the webs had to be highly inclined to support the deck [5]. As vertical webs carry shear transfer best, the existing standards did not allow verifying structural integrity. Hence, formulas from the latest research were used instead [6], and with their help, it was possible to provide theoretical proof of the integrity. However, a full-scale test was required as an empirical verification [41]. The highly inclined webs of the D6 segments made it possible to strengthen the cross-section with transversal post-tensioning running through the bottom of the webs as well as the top [7]. All this determined the overall design of the D6 segment [8]. The width of the segment also required a new spatial layout of the longitudinal post-tensioning [9]. When match casting segments, hydration heat deforms the segment cast the previous day, and the fresh concrete sets against this deformation, resulting in gaps between the segments. Engineers call this phenomenon the "bowing effect" (Brockmann 2001). Formulas gave a value of length $b/height\ d < 9$ as the maximum for segmental construction. A disbelief in the soundness of existing knowledge and careful consideration of the problem lead to the conclusion that match casting would be possible for D6 segments ($b/d = 27.2/2.55 = 10.7 > 9$) [10]. Again, the full-scale test provided the required proof.

In areas of the ramps and toll plazas, a widening of the cross-section was unavoidable. Thus, the design integrated the mainline piers and the portals with two conventional columns on the outside [11]. The contractor decided to again use segmental construction for these portals after signing the contract [12]. The final pile foundations were equally determined after signature, and they were assembled from a number of spun piles of 25 m length that could be welded together to reach the required depth of up to 50 m [13]. A design firm designed the rebar cage for the pile cap with a rebar content of $160\ kg\ m^{-3}$. This was unacceptable to the design/build contractor who reduced it to $120\ kg\ m^{-3}$ by making efficient use of existing standards [14].

During the tender stage, a more conventional design with two smaller segments was the starting point of deliberations. However, perceived market pressure induced the contractor to push the designer toward a more economical design. The original design firm was not able to improve the design sufficiently, so the contractor addressed Jean Muller International for a more innovative solution, which became the D6 segment design. When signing the contract, a general design idea existed and quantity take-offs were possible. However, there were major risks associated with the design, as a full-scale test span could only provide proof of the validity of the assumptions later. The design/build contractor determined the price based on his design experience, not on checked and cross-checked design. He was also driving the development of the design. Jean Muller International proved to be highly innovative in conjunction with the contractor. The owner accepted the design with few comments. The test span was not part of the contract, but its use was a decision of the project management with a price tag of 1 million USD.

Subcontractors assembled the toll collection and traffic control systems from standard parts. However, the software running the systems is largely an innovative application, as the subcontractors had to take into account the specific boundary conditions for the project [15, 16].

There were many strong cross-impacts between the system design and the chosen construction technology, as both are intricately related. Only the interplay of technological competence (mostly the contractor) and design competence (mostly the lead designer) made the solution possible.

13.1.2.2 Construction Technology Innovations

In the following, I will analyze 17 innovations to the construction technology (Table 13.2). Again, many of them are interdependent and influence the design or, in turn, the design influenced them. The width of the bridge prohibited the use of the more typical overhead girders for segmental construction. The chosen underslung girder influenced the top of the columns (opening between arms, inclined bearings, no crossbeam) and the form of opening between the upper arms of the column affected the girder [17].

The swivel crane was an integral part of the girder and could be placed at the end of the girder or on top of the erected bridge. This allowed for delivery of the segments either at grade or on the bridge. Plans were to erect generally from the bridge deck [18]. The swivel crane lifted the segments from the trailers and placed them on chassis that transported the segment to the other end of the girder [19]. The last segment placed was the pier segment close to the swivel crane. Next, crews inserted and stressed the longitudinal post-tensioning. After this activity, the segments acted as one span and it became possible to lower the span onto the bearings. Finally, the girder was launched into the next span. Due to the design constraints, the piers were unique, a fact that is also true for the formwork [20].

As the D6 segments were innovative, it follows that the same holds true for the molds in which they were produced [21]. Specialized software guided the surveying of the molds for segment production [22]. The schedule assumed a 2-day cycle for the erection of a D6 span. This doubled previous erection speed with regard to time per square meter bridge deck [23] and became necessary because of the short design and construction time of 42 months. It entailed consequences for the required amount of main resources: molds, storage area, trailers for transportation [24], and number of girders. The contractor ordered all of these expressly for the project as special designs. At the same time, the available resources predetermined the production sequence in the yard. First crews prefabricated the struts and prepared the transversal post-tensioning and deviator blocks for the rebar cage. Rebar workers prefabricated the rebar cage and tower cranes lifted it into the surveyed molds. The concrete crews mixed the concrete in two batching plants and transported it with the help of truck mixers and conveyor belts. P/T crews applied partial post-tensioning of the transversal tendons in the molds before moving them to the match cast position by rail and, a day later, to curing, also by rail. Shuttle lifts brought the segments to the storage area [25] and double-stacked them to save space [26]. This is a description of a production line similar to what we find in the car industry: the product (D6 segments) moves while workers and tools are stationary [27]. The whole yard acted as a consignment center in logistics (Sullivan et al. 2010): Most deliveries arrived at the yard, and logistic crews dispatched them from there to the site [28].

Using the same construction technology for the production and erection of the portal beams was a refinement initiated and designed on-site [29]. It required the design of a completely new launching and erection girder: as the portals were perpendicular to the mainline, the erection girder sat on a turntable on top of the launching girder and it was possible to turn it by 90° [30]. Project engineers also developed lifting devices for the D6 girders on site. Initially, they planned on using two large mobile cranes. Due to a monopoly situation however, the rental fee for the mobile cranes was unacceptable and the lifting devices provided a cheaper and more flexible solution [31].

The girder crews achieved the assumed 2-day cycle for a first time after six spans. At the end of the project, one crew out of the five D6 girder crews had increased the speed to 1 day per span (i.e. quadrupling the previous erection speed!). The work crew achieved this through a string of incremental innovations, which they devised themselves [23].

At the same time, the project size made the investment in other owned equipment more economical than renting (25 mobile cranes, 8 tower cranes, 10 truck mixers, and 2 batch plants) [32]. Construction management chose five piling rigs were especially equipped for the conditions of the project [33].

13.1.2.3 Innovations Within the Technical Organization

The technical organization arranges the integration and flow of technical components. It is possible to identify eight innovations in this area (Table 13.3). The sheer size of the project required an innovative approach while at the same time making it possible in the first place. Most of the innovations described thus far had to be integrated with the idea to optimize the system and not the subsystems. Such an approach will often result in suboptimal solutions for subsystems. In construction, nobody had applied this principle on a similar scale [34]. The design/build megaproject contract [35] offered the chance to integrate design and technology as described in the previous chapters. The innovative result rested on three foundations:

- A contractor with strong design and construction experience (not only a management contractor) and with a capability to estimate the financial and temporal consequences of a journey into the unknown
- An experienced and innovation embracing design firm
- A welcoming environment with a local joint venture partner and an owner who grasped the opportunities. Till date, the BangNa Expressway has operated for more than 20 years without any technical problems.

The size of the project with the production of 22 000 similar D6 segments at its core necessitated and allowed for the production flow [36] and the implementation of the world's largest precast yard and operation [37]. It also provided the funds for the investment in tailor-made equipment. While construction often operates under the conditions of a substitutional production function (e.g. labor can be replaced by equipment and vice versa), the BangNa expressway used a limitational production function [38] with a fixed ratio of inputs (five D6 girders, 22 trailers, 1650 storage places, 48 molds, and six overhead D3 girders with similar support). Adding more labor or another 10 trailers would not have increased construction speed. The result was a very high construction speed, albeit at the cost of reduced flexibility. Altogether, 11 girders had to be coordinated (five D6 girders and six D3 girders for ramps and toll plazas) [39]. Rebars were cut and bent in the same yard where concrete was mixed, and both were used in the yard or delivered to the site. This is a vertical integration seldom found in construction, and it gave the possibilities to technically integrate and optimize to a high degree [40]. Two spin-off companies provided more vertical integration, a spun pile company producing the 900 km of spun piles for the project and additionally serving the market and a post-tensioning company that produced the post-tensioning system for the project [49, 50].

Size rendered the full-scale test span feasible, and the innovative design assumptions made it necessary. It proved that the system performed as expected. Thus, the design analysis was tested and found to be correct [41].

13.1.2.4 Innovations Within the Management Organization

Eleven innovations are of interest in the management organization (Table 13.4).

The amount of work within the given time – the density of work – was unprecedented [42]. The same holds true for the resources; 500 employees and 5000 workers strived to complete the project on time. The ICJV recruited, trained, and released them within 42 months of contract time plus 11 months of time extension [43]. The make-or-buy decisions were mostly in favor of a make. The contractor used subcontracts mainly for electrical and mechanical works [44]. On two previous bridge projects, the partner companies organized project management, construction management, and yard management in three different joint ventures. Due to the size of the BangNa Expressway, they integrated, for the first time, all these functions in one joint venture [45]. Consequently, the internal management made ample use of decentralization of management functions [46]. Project management decided to organize the largest part, construction management, as a matrix organization, with area and technology functions separated [47]. Design management had to integrate the work of firms from eight countries across three continents [48]. I have already mentioned the start-up companies for spun piles and post-tensioning within the technical organization [49, 50].

Commissioning, handover, and opening for traffic was a staged process in eight phases [51]. On the site of the BangNa Expressway, two owners had contracted work for execution in the same place and at the same time without coordinating them. While the Expressway and Rapid Transit Authority (ETA) of Thailand was the owner of the BangNa Expressway, the Department of Highways (DoH) was the owner of Highway 34 and the right-of-way. The latter had signed a contract for the construction of lime piles underneath Highway 34. The piling rigs interfered with the expressway construction, and DoH gave them precedence without handing over a schedule of works. Consequently, the design/build contractor of the BangNa Expressway undertook the coordination without contractual obligation [52].

13.1.2.5 Innovations Within the Legal Organization

Six innovations are evident within the legal organization. Three were premeditated and another three developed during the execution of the project (Table 13.5).

According to the chosen definition, all of these innovations led to an improved input/output relationship. The design/build/finance contract with the project owner (ETA) was a new type of contract for the contractor [53]. He signed an independent financing contract with four Thai banks [54] and aligned it with the design/build/finance contract. ETA compensated the contractor with promissory notes issued by the government, and the banks reimbursed the stated amounts to the contractor. At the end of a construction phase, ETA honored the promissory notes and the contractor repaid the banks [55]. The parties to the design/build/ finance contract introduced eight construction phases in order to minimize outstanding payments.

The contractor finished construction in 2000 and handed the project over to the owner for operation. However, only 15 years later, the joint venture dissolved [56]. Only then had the contractor solved all open points of the contract. The uncoordinated works of ETA and DoH resulted in a claim of 280 million USD. A FIDIC design/build contract was the basis for the design/build/finance contract and stipulated arbitration as the final instance for claims. The arbitration committee conceded this amount to the contractor [57]. However, the Thai government took the case to the Thai Supreme Court. The Supreme Court decided against the contractor and declared the contract illegal. The BangNa Expressway thus holds – besides all other records – the distinction to be the longest illegally built bridge in the world [58]. The expressway has served the Thai public in phases since 1998 and continues to do so without fail and quality problems.

13.1.3 Conclusions and Implications

I would like to draw some general conclusions from the BangNa Expressway as it is a paradigmatic megaproject.

13.1.3.1 Megaprojects are Innovative

The BangNa Expressway provides overwhelming evidence of innovation in construction megaprojects. This is supported by secondary data collected through interviews over 25 years at the Western Bridge and the Eastern Rail Tunnel of the Great Belt Crossing in Denmark (1988–1998, construction cost about 1 billion USD each), the Taiwan High Speed Railway (1998–2007, about 18 billion USD), and the Qatar Integrated Railway System (2008–2022, 40 billion USD). Acknowledging the fact that there are also construction projects without a single innovative idea, we can conclude that we must characterize the construction industry by a continuum of non-innovative to highly innovative, depending on the constructed product and constraints such as the type of contract and ability of the contractor.

The discussion showed that the complexity of the project created ample opportunities for innovation, and this leads to the assumption that perhaps project complexity (vs. project simplicity) influences innovation more than other industry characteristics.

13.1.3.2 Strings of Incremental Innovations

It is truly astonishing that a crew of Thai workers and expatriate supervisors were able to double the world-record construction speed from a 2-day to a 1-day cycle. The complexity of the erection process allowed for a string of incremental innovations bringing about the result. There is no way to trace these incremental innovations systematically. They remain hidden even within a longitudinal study.

13.1.3.3 Innovation in Megaprojects is Systemic

To carve out single innovations has been a cumbersome task. This is only marginally due to deficient definitions as they are not operational. More salient is the fact that the described innovations are mostly systemic, with strong cross-impacts. Product, process, and organization form different subsystems of the innovations. The most suitable characterization of the BangNa Expressway is that it is an innovative project. Only to arrive at this conclusion, it is worthwhile to analyze the overall innovation and unpack it. There are no clusters of innovation in the BangNa Expressway; there is a system of innovation.

13.1.3.4 Innovation is not Necessarily Beneficial to All Parties

Innovation as a nontrivial change can be positive or negative. A positive example of the BangNa Expressway is the development of the D6 segment; a negative one is the settlement before the Thai Supreme Court. Viewing innovation indiscriminately as something positive is naïve (Sexton and Barrett 2003). As economists distinguish between pros and cons, definitions of innovation need to discriminate between the types of outcomes. The chosen definition of "an improved input/output relationship" provides such a discrimination principle, albeit one that is not easily operational. The decision by the Thai Supreme Court provided an improvement for the owner, not for the contractor.

13.1.3.5 Contractors Can Manage Single-Project Innovations in Megaprojects into Good Currency

Gambatese and Hallowell (2011) use as a definition for innovation "actual use of a nontrivial change," and although the definition does not provide for it, they claim that diffusion is also required to provide validation of the change. Based on a definition of "management of new ideas into good currency," Winch (1998) also adds the use of the innovation in other projects as a criterion. The BangNa Expressway case study shows that this view is not only an addition to the basic definitions provided but also that it necessarily applies to megaprojects. The size of the BangNa Expressway surpassed that of the 11 long span and multi-segmental bridges in the sample of Slaughter and Shimizu (2000) by 50%. The contractor definitely managed the innovations into good currency in a single project, validated by the full-scale test, $1\,900\,000\,m^2$ of bridge deck and 20 years of operation: Construction innovations do not need to be diffused to become innovations. Authors who dismiss this idea should provide a consistent definition of innovation and take into account that sometimes the one-off nature of construction projects (especially of megaprojects) does not allow for diffusion. Otherwise, such construction projects would become non-innovative by definition.

13.1.3.6 Innovation Champions Act on All Hierarchical Levels

The idea of champions for innovation as promoted by Nam and Tatum (1997) does not imply that top management officials need to be superheroes with power to push innovations through a reluctant organization. Instead, technological competence, power, and resources are required. Innovation champions exert leadership. Leadership, however, is a cultural concept (Hofstede and Hofstede 2005). The BangNa Expressway was a German-Thai co-management, and Germans as well as Thais are more collectively orientated than Americans (the sample in the study from Nam and Tatum). Whereas leadership is a convincing narrative in the United States and as such effective, this does not hold true to the same degree in other cultures. The described innovations arose on all professional levels: top executives of the companies and top executives of the project; middle managers in the companies and middle managers in the project; finally, workers on the project (doubling the speed by incremental innovations). Often – and in all cases for the contractor's top management level – the decision was a group decision. One more final observation: the tables also permit distinguishing between decisions taken before signing the contract (TM-pre, Bid) and afterwards. The analysis yields the result that the design/build contractor accepted many risks. This does not support viewing the whole construction industry as conservative.

13.2 The Innovation Process

As described in Section 13.1, construction megaprojects often exhibit high degrees of innovation due to, amongst others, their singularity and size. They are captivating, and we must expect the innovation process to be captivating as well. Since little information is available, descriptive research on the innovation process by contractors might benefit a better understanding of the process in the industry. Longitudinal studies again promise to generate the deepest insight possible because innovating takes time. The following data come from four projects. They reveal a highly complex, iterative, and messy process with nine overlapping process groups. The contractors were never able to define problems completely and used simplifying heuristics. Satisficing behavior and bounded rationality replaced optimization. I also found that the innovation process takes the form of problem-solving, and this opens new insights to the old topic of construction industry characteristics: it reveals that, in megaprojects, neither traditions nor adversity to change prevail; instead, pit bull terriers of innovation are on the loose.

13.2.1 Introduction

Although construction megaprojects often show a large degree of innovation, we have no rich model of the innovation process that leads to this result. Such a model would detail the innovation process in more than just a few consecutive stages. It would have to give answers to the why, how, where, what, when, and who of innovation. A rich model should not only provide understanding for the scholar but also enable managers to guide and influence the process.

How can we establish such a model? I believe it is preferable to proceed by describing the decisions and actions of the people involved. Experienced managers and engineers plan and execute megaprojects, and they create best practices by trial and error. An innovation process model must reflect these best practices. Results will show that a successful innovation process for construction megaprojects is not a journey (Van de Ven et al. 2008) but a very bumpy high-speed drive into the unknown with pit bulls in command who relentlessly push for a large degree of innovation.

Since the innovation process in itself is a bumpy ride, it seems preferable for the development of the argument not to follow suit. Therefore, I will develop the narrative in this sequence: (i) Some reflections on different ways to generate innovation are as necessary as a definition to plant our ideas firmly in a construction context. (ii) An analysis of existing innovation process models will serve as reference points to develop a rich model. Results on barriers to innovation in construction are abundant. Consideration of such barriers helps choose cases that provide many innovative examples. (iii) After choosing promising cases, we need to develop a research strategy that allows pinpointing innovations and retracing the process to its origin. (iv) All these preparatory steps permit collecting and presenting raw data from the research. (v) From the raw data, I will extract the rich innovation process model. (vi) Consequently, two examples shall help understand the application of the model. (vii) The conclusion provides space for a critical evaluation.

An innovation model is part of the implicit knowledge of the project participants. Asking direct questions in the form of closed interviews or surveys can only tap into the top-of-the-mind associations; this will not capture the full picture of what happens. Only long-term observation can achieve this goal. To make the model "general" in a certain way, one case study is not enough; therefore, I have conducted four (Table 13.7). Feedback from practitioners allow calling the model in a preliminary way "general."

13.2.2 Approaches to Generate Innovation and Definition

The distinction between basic and applied research is well-known, along with the relevant goals for each endeavor. Basic research tries to develop theory and applied research to find solutions to problems. Different means are at our disposal for basic as well as applied research. Paradigmatic is rationality (Popper 2002), but also possible are emergent solutions (Mintzberg and Waters 1985) or serendipity (Hamel 2002; Rosenthal 2005; Loosemore 2014). Companies organize basic research, most often in research and development (R&D) departments and applied research in project groups.

Companies follow the goal of profit maximization whether they pursue basic or applied research. Rosenberg (1990, p. 169) summarizes the motives of private companies to engage in research: "*Thus, it is doubtful that business decision-makers often sit down and ask, in an abstract way: Should we do basic research? How much basic research should we do? Obviously, private firms feel no obligation to advance the frontiers of basic science as such. Presumably, they are always asking themselves how they can make the most profitable rate of return on their investment.*"

Turning attention to construction megaprojects, we will notice their almost absolute singularity. Task, social, cultural, operative, and cognitive complexities are at the limits of human understanding and differ largely from megaproject to megaproject. In such cases, basic research will not increase the return on investment because benefits are limited in scope. Singularity distinguishes megaprojects from mass production in pharmaceutical or chemical industries where innovative products sell millions of times. While the return on investment in construction seldom benefits from basic research, it might benefit from applied research generating new solutions.

If the conclusions above are correct and project teams generate applied innovations for megaprojects, we also have an explanation why they are not traceable through accounting systems. They are not booked into an account "R&D," but into one with the name of the megaproject under consideration.

There is no shortage of definitions for the term "innovation," and they serve different purposes (Baregheh et al. 2009). In the context of an innovation process for megaprojects, such a definition must be process-oriented, encompassing, and business-oriented. I have provided one in Section 13.1, which also applies to the process model.

13.2.3 Innovation Process Models and Barriers to Innovation

The academic literature on innovation is extensive. For my purpose, it is sufficient to discuss four stage-based models of the innovation process as well as barriers to innovation in construction. Zaltman et al. (1973) and Van de Ven et al. (2008) have described the innovation process in general; Tatum (1987) and Laborde and Sanvido (1994) have done the same for the construction industry. The models are stage-based and shown in Table 13.6. I have arranged the stages in a way that comparable ones line up.

A comparison of the four staged-based processes shows some similarities. However, it seems that all neglect the idea of profit-making in the business environment. The disregard of the business side violates the well-known definition by Van de Ven (managing ideas into good currency). Also missing in all the models is information on how ideas are developed, who the drivers of innovation are, what heuristics these drivers use in the problem-solving process, and how they make decisions with regard to the innovation. While the models are rather simple, most texts describe the processes in detail, especially Van de Ven et al. (2008) who expound the important idea of how periods of divergence and convergence alternate during the innovation journey.

The models of Zaltman et al. (1973) and Van de Ven et al. (2008) describe innovations on the firm level. This is also the focus of Tatum's model of innovation in construction (1987).

Table 13.6 Innovation process models.

Zaltman et al. (1973)	Van de Ven et al. (2008)	Tatum (1987)	Laborde and Sanvido (1994)
Knowledge awareness	Initiation	Recognize forces and opportunities	Identification
Formation of attitudes	—	Create climate for innovation	—
—	—	Develop necessary capabilities	—
Decision	Development	Provide new construction technology	Evaluation
Initial implementation	Implementation	Experiment and refine	Implementation
Sustained implementation	—	Implement on projects and in the firm	Feedback

Process models, such as that of Laborde and Sanvido (1994), are so general that they can describe almost any process; it bears much resemblance with the basic plan/do/check/act model.

Arguably, such models have little explanatory power for construction projects. What is missing is a model for Tatum's stage "Provide new construction technology." On-site production forces the construction industry into a predominant project organization. Innovation in these construction projects is, therefore, the most relevant, and we have no model describing the process. Four idiosyncrasies concerning innovation characterize the construction industry and set it apart from manufacturing: (i) *Project-based innovation*: Contractors spend little money on formal R&D (Laborde and Sanvido 1994), and the vast majority of innovations are project-based (Brockmann et al. 2016). (ii) *Awareness of innovation opportunities*: This is a prerequisite when applying for the tender documents of a megaproject. Most of the time, a successful prequalification is required to be eligible for tendering. The owner triggers awareness from the outside; the innovating company in manufacturing must generate it from the inside (Van de Ven et al. 2008). (iii) *Focus of innovation*: Regardless of the procurement method chosen by the owner, all tender documents define, to some degree, the final product (building and structure). These documents limit and guide the innovativeness of contractors. In manufacturing, no such constraints exist (Van de Ven et al. 2008). (iv) *Time pressure for innovation*: The owner sets another limitation by the tender and construction schedules. Contractors have two main periods during which they can innovate: during tendering and when mobilizing. The contract price reflects innovations found during the tender period; further innovations during the mobilization period can improve the profitability of the project. Each period lasts between 6 and 12 months for megaprojects (according to our data), which is a very short time in comparison to the innovation journeys described by Van de Ven et al. (2008), where one process lasted 21 years. Of course, it is also possible to innovate concurrently with the implementation, but during this time, innovation can never be the focus. I conclude that the existing innovation process models do not appropriately address these idiosyncrasies. Therefore, I will try to answer the basic questions of who, when, where, what, how, and why of the innovation process in construction megaprojects.

Stages as sequential units along a time axis often do not reflect the nature of a process adequately. Process groups capture the reality much better: they facilitate the description of processes in which more than one activity is going on at a given moment in time.

Various authors discuss barriers that confine innovation in construction. I have discussed this already in Section 13.1. To discover the innovation process, we need projects that avoid as many barriers as possible. Most barriers do not affect megaprojects to a large degree. The four chosen projects (Table 13.7) allow for comparison between different contracts, different supply chains, different competencies, different owners, and different regulations. It will become evident that barriers affect the degree of innovativeness but not the process.

13.2.4 Data Summary

The following discussion presents the main answers to the questions of why, how, where, what, when, and who regarding the innovation process in megaprojects.

Table 13.7 Research cases.

Case	Location/time	Organization/ (delivery method)	Predominant methods
Great Belt Rail Tunnel	Denmark (1988–1997)	Contractor (Design/bid/build)	Interviews
BangNa Expressway	Thailand (1995–2000)	Contractor (Design/build)	Action research
Taiwan High Speed Railway	Taiwan (1998–2007)	Contractor (Design/build)	Interviews
Qatar Integrated Railway Project	Qatar	Owner	Action research, interviews

Why do contractors engage in innovation? There were two major reasons for engaging in innovating – to increase competitiveness and profitability. Contractors can only increase their competitiveness during the tender period. How much of the monetary gains from innovation is used for this purpose depends on the appraisal of the market. Project teams have a short-term focus on their project; for them, winning the tender and generating a profit is paramount.

How? Formal R&D played no role in the four projects; project-based research was strong. Most innovations derived from problem-solving; emergent solutions played a role, but serendipity did not.

Where? People in the head office created the largest amount of innovations during the tendering period (between projects); project managers and engineers developed fewer innovations during mobilization and an even lesser amount during implementation (within the project).

What? To describe the types of innovations generated, the BangNa Expressway can serve as a model: it features 58 identified major innovations (Section 13.1).

When? Answering this question is at the core of our research and we will explain and exemplify this later. People drive the innovation process. This leads to the question: Who is doing what? We will discuss this along with the model after the following preliminary observations.

Who? During the tender period, bid teams work out the details of the bid including innovations. Decision-makers are periodically involved in the process and might push for more innovative solutions or limit risk taking. After signing the contract, a project team takes responsibility for the implementation. Not all innovations look feasible to them, or they might sponsor additional ones. In the data, I could not clearly detect champions of innovation belonging to one single group. The pit bull metaphor describes behavior instead; as an innovation metaphor, it conveys the idea of a relentless, aggressive pursuit of innovative solutions. As the bid team is head over in the development of solutions, they might fall into the trap of groupthink (Janis 1972). The decision-makers and the implementers can, in such a case, counter-balance optimism bias (Flyvbjerg et al. 2003) with caution and rein in the pit bulls. However, in the case of the BangNa Expressway, the decision-makers were as optimistic as the bid team. They were not satisfied with the design of the associated consultant and switched to another more creative design firm pushing for innovations. In the same case, the project team refused to implement two innovations while coming up with their own innovations during mobilization and construction.

Bid teams for megaprojects are large and have access to considerable amounts of money, facilitating innovative solutions. The four bidding teams for a recent megaproject, the Femern Belt project in Denmark, reportedly spent 20 million euros each for the bid preparation between 2014 and 2015.

13.2.5 The General Model of the Innovation Process

The general model of the innovation process consists of two parts: (i) project planning as a circular activity with 11 partial plans, and (ii) nine partially overlapping process groups. Both the project planning activities and the project phases emerged from the collected data, and I will explain them later.

13.2.5.1 Project Planning

Project planning is an activity that is indispensable for all construction projects (Section 8.3). This is the domain where contractors can always innovate, regardless of the procurement method. The information generated is required to submit a bid. While there is no way around project planning, the detailing depends on the type, size, and complexity of the project. The estimator performs project planning for small and mid-size projects. A team of specialists as part of the bid team is crucial for preparing a bid for large-scale projects. Megaprojects are, without exception, singular, and the differences, for example, between tunnel or bridge projects or even between two bridge projects, are substantial. Consequently, the team of specialists must develop a novel solution in accordance with the boundary conditions of the focal project: project planning for megaprojects is an innovative activity by necessity.

It is a demanding task for megaprojects – messy, non-sequential, and frequently intuitive. During project planning, the pit bull terriers in the bid team run around in circles, looking for the best attainable results without rest. It is impressive to watch the specialist team pursue solutions that lead to a dead end and start again without tiring. It is not a task for those who are easily discouraged. The teams carry out project planning during the tender stage and perfect it during project execution.

When analyzing project planning in megaprojects, we must recognize that it has nothing to do with rationally sequential work. Van de Ven (2008 p. 16) describes the process adequately – albeit with regard to innovation: "…*the innovation journey is a nonlinear cycle of divergent and convergent activities that may repeat over time and at different organizational levels if resources are obtained to renew the cycle.*" The difference between this description and project planning for megaprojects is the time constraint. At the submission date, the contractor must present a solution and a price. Once a company decides to prepare a bid, it also must appropriate the necessary resources.

To understand the complexity of the system of partial plans in Figure 8.19, the data from the BangNa Expressway may serve as illustration: the 11 partial plans have 55 possible interrelations causing cross-impacts. The estimate had more than 10 000 items with four cost types each (labor, material, equipment, and subcontracts). The schedule had more than

2000 activities. The site installation had a value of 150 million USD. The value of the equipment was around 50 million USD. Human resources amounted to 500 employees and 5000 workers. People worked along a stretch of 55 km. It is not possible to express the technical, management, and legal organization in numbers. The large task complexity of a megaproject makes it clear that expert judgment often replaces rational analysis. Once the contractor signs a contract and the project team takes over, the task complexity increases manifold as the team starts detailing all plans.

Returning to the 11 partial plans and remembering profit maximization as the paramount goal, we can formulate the task for a contractor: minimize the costs as they depend on the other partial plans.

13.2.5.2 Nine Partially Overlapping Process Groups

The project members and interviewees did not describe an innovation process. However, from observation by action research and some information from interviews, I was able to establish nine process groups of a typical innovation process. These are (i) awareness of a problem, (ii) analysis of the problem, (iii) searching the set of known solutions, (iv) determining the gap, (v) what-if questions, (vi) satisficing, (vii) adequacy of resources, (viii) organizational decision-making, and (ix) implementation.

- Process group 1: Awareness of a problem
  With regard to megaprojects, the "problem" starts with the decision to submit a bid. Companies that engage in megaprojects are well-aware of the consequences and the resources required. The bidding phase will typically cost several million USD. Depending on the delivery method, the contractor needs to find a solution for the problems of project planning (design/bid/build) or for the design and project planning (design/build and engineering/procurement/construction (EPC)). The focus is on the overall solution for the new project. Finding the best possible overall solution means accepting suboptimal solutions for some of the partial plans given the time constraints in construction innovation and the complexity of the problems. Figure 13.4 shows problem awareness as

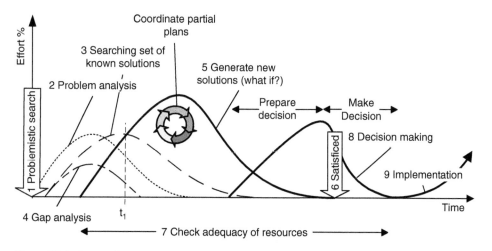

Figure 13.4 Innovation process.

a singular impact (arrow). Yet, in reality, this is often a lengthy process, including expression of interest, information gathering, prequalification, screening of the documents, and finally, commitment to the project. However, the commitment happens at a given time.

- Process group 2: Analysis of the problem
Megaprojects are complex, and the information given or required is overwhelming. As the contractor tries to understand the design (conceptual design or detailed design), additional information is most often of vital importance. The analysis is not always a rational process as time pressures necessitate taking shortcuts and this involves prioritizing. In our data, there exist many examples of problems, which the contractor did not understand at a point in time. The following process groups force the project team to repeat certain parts of the problem analysis to generate a thorough understanding: developing a solution and analyzing the problem overlap.

- Process group 3: Searching the set of known solutions
The search criteria are not clear and, therefore, the search process is ill-defined. In the case studies, proprietary knowledge databases with data on past megaprojects did not exist. The reason for this might be the novelty of megaprojects and the lack of commonalities between megaprojects. Informal brainstorming or discussion groups are typical ways of accessing previous project knowledge. While managers and engineers develop solutions, they also remember solutions from previous projects and use them in problem-solving.

- Process group 4: Analysis of the gap
Because of the singularity of megaprojects, a large gap between previous projects and the focal one is inevitable. Most of the time, the transfer of previously generated solutions will provide highly unsatisfactory results, whereas at other times, it is simply impossible. The gap analysis lays open the degree to which the focal megaproject is novel. Again, the analysis is not detailed. Due to the complexity of the megaproject, there simply is not enough time at this stage to describe the gap completely.

- Process group 5: What-if questions
When the known solutions are inadequate, then what-if questions are a simple heuristic to generate new solutions. Pursuing this constant questioning long enough is a very powerful tool to develop innovations. The data provide very few and only minor examples of a genial solution, presented more or less unexpectedly. It is a tedious work process, in which managers and engineers carve out the innovations gradually. Part of this work is the constant balancing of often contradicting demands between the partial plans. The degree of innovation depends on the resolve of the bid team and later the project team to keep asking what-if questions and going into the next round of adjustments to the partial plans. Here again, I observed people showing the doggedness of pit bulls.

- Process group 6: Satisficing
Innovators constantly used satisficing to end the search when looking for innovations. The complexity of megaprojects does not allow defining a finite set of variables that describe a given problem (Li and Love 1998). From this follows the inability to clearly structure and analyze the problem. Some of the consequences from decision-making will remain unknown. In addition, the number of possible solutions is infinite. Therefore, the innovators in megaprojects face three impossibilities: (i) the impossibility to determine all variables defining the problem, (ii) the impossibility to structure the

problem completely, and (iii) the impossibility to gauge the sets of alternatives and consequences. When a problem is ill-structured, the solution cannot be completely rational; the innovators must also rely on intuition. They must content themselves with a solution that looks good enough to themselves. This is satisficing behavior. An owner, on the other hand, should demand solutions from different competitors in order to choose the optimum offer among them by rational analysis.

- Process group 7: Adequacy of resources
A newly developed solution must be buildable. This is what differentiates an invention from an innovation. For a contractor, the question is whether he has command of the required resources. Financial strength is a basic requirement for megaprojects but more crucial is the human capital of a company: Does it have access to the right people who can implement the inventions? Is there enough knowledge within the company? Is the required learning process possible? The technology space describes the technical knowledge base of a company. Given an invention, the contractor must be able to expand his technology space accordingly; it is necessarily a step into the unknown. Sometimes, this is a large step requiring a lot of good judgment and courage – the courage of pit bulls. For this, the ability to learn assumes paramount importance.

- Process group 8: Organizational decision-making
Organizational decision-making is at the core of organizational theory. Task ambiguity and bounded rationality characterize decision-making for innovations in megaprojects. A sizeable amount of literature exists on decision-making in organizations within behavioral organization theory. Influential are the works of March (1988) and Cyert and March (1992). Social psychology covers another aspect of decision-making, e.g. the phenomenon of groupthink (Janis 1972). Also of great interest are the contributions from cognitive science (Kahneman 2011). Individuals or groups take decisions throughout all stages of the innovation process. Organizational decision-making in this context is the process of agreeing to or rejecting an innovative solution by those who carry the business responsibility.
A typical way of coming to a decision is in a meeting with a presentation of the problem, alternatives, and the proposed solution with its pros and cons. Data from observation show that informal groups with members of the bid team and decision-makers often prepare the decision in face-to-face communication. As ambiguity pervades the process, rationality is only one aspect and personal preference is another; risk attitudes play an important role. There are also several instances of groupthink with negative outcomes in the dataset. In consequence, the main decision-making meeting is prepared and followed-up; it is a lengthy process. During this process, the proposed solution might find no acceptance, which requires returning to some previous process group.

- Process group 9: Implementation
There is a clear difference in the risk appetite between the bidding team and the project team after contract signature. The latter is responsible for successful implementation of the inventions and turning them into innovations. We observed that the project team rejected some inventions or parts thereof. This must not necessarily be disadvantageous as it might regulate overambitious inventions. It is a twofold approach, first with an emphasis on the desirable, then on the achievable. The project teams never rejected an important part of an invention; differences only arose on the fringes.

Project participants did not describe the different steps in the innovation process as phases but as overlapping process groups (Figure 13.4). The arduous process of integrating the partial plans is part of the development of the new solution by asking what-if questions. The sequence of overlapping process groups in Figure 13.4 is again a simplification. There were several decision meetings in all researched projects. Thus, the participants went through several iterations for most process groups.

In the next two chapters, I present two examples from the four research projects (Great Belt Tunnel, BangNa Expressway, Taiwan High Speed Railway, and Qatar Integrated Railway Project) to illustrate the described general process. First, I will detail the innovation process for the overall design and construction of the BangNa Expressway in Thailand. This was a design/build/finance contract where the contractor had to develop product as well as process innovations.

The second example comes from data on the Great Belt Tunnel in Denmark and describes the process for the development of the site installation. This is a design/bid/build project with a detailed design provided by the owner. The contractor only had the possibility to come up with process innovations for the implementation of the owner's design. In the first case, the contractor developed product and process innovations simultaneously, and, in the second case, the process innovations were both limited by and subordinated to the given product design.

13.2.6 Product and Process Innovations for the BangNa Expressway

- *Awareness of a problem*: Prior to the BangNa Expressway project, two construction companies had contracted two other elevated expressway projects with the same client in Bangkok using precast segmental bridge technology. A third had built an elevated expressway, also in Bangkok, using a different technology, i.e. pre-stressed girders and cast in-situ decks. The American Association of State Highway and Transportation Officials (AASHTO) standardizes such girders. The experiences gained alerted the companies to the problems of the new project.
- *Analysis of the problem*: A total of 18 months passed from the decision to prepare a bid until contract signature between the owner and the contractor. During that time, the bid team comprised up to 20 engineers. There was sufficient time to appreciate what it takes to design and build the longest bridge in the world within only 42 months.
- *Searching the set of known solutions*: With the experience of three expressways in Bangkok, the bid team naturally considered previous solutions. The two construction technologies were compared (segmental construction versus AASHTO girders with in-situ deck), and their suitability for the new project was considered.
- *Analysis of the gap*: In comparison to the previous projects, there were some notable differences. The size of the BangNa Expressway was four times bigger than that of any previous project expressed in terms of deck area. The bridge gradient was higher – 18 m instead of 12 m. The cross-section comprised six lanes instead of two times three separated lanes; however, the expressway built with AASHTO Girders had six lanes. With 42 months, the time for design and construction was much shorter compared with the other projects.
- *What-if questions*: The bid team discussed the two basic alternatives thoroughly. In the end, the group chose precast segmental construction because of the predicted lower price

and faster construction progress. The design at that stage showed two bridge superstructures side by side on a central pier using D3 segments developed for the previously built expressways with a maximum of three lanes. The bid team considered the solution unsatisfactory with regard to construction speed and price. They assumed that the price would not be competitive enough and began another search for a faster and cheaper solution. The most important what-if question centered on the idea of developing a single segment for six lanes (D6 segment) with a total width of 27.20 m. This would almost double the previous maximum width of the D3 segments (15.60 m) and the erection speed. Among the options discussed were designs with three, two, or one central web in addition to the inclined outer webs. This change entailed switching design firms. Figure 8.5 shows the final option in the upper part. It includes two internal struts and two highly inclined webs. This is a concrete example of pit bull aggressiveness with regard to innovation and the market. The comparison of the two options in Figure 8.5 illustrates the idea that the contractor could achieve the two main objectives – increased construction speed and lower cost – with the D6 segments. The D6 segment design was revolutionary, as nobody had built anything similar before. Accordingly, this option entailed a number of unresolved problems (Section 13.1).

- *Satisficing*: The development of a design option for the BangNa Expressway was a team process, and it stopped when the team felt it had surpassed its aspiration level. The two groups within the team, one favoring precast segmental construction and the other pre-stressed AASHTO girders, did not reach an agreement at this stage. They reached agreement only during the final decision-making meeting.
- *Adequacy of resources*: Decision-makers can merely appraise the likelihood of implementation. Only during implementation will they find out how difficult and costly it is. However, they must make a decision before submitting an offer, and this constitutes a leap of confidence for contractors. Although other questions were raised when developing the solution, in this particular case, the most important one was whether one span could be erected in 2 days, for the entire schedule depended on this point. Just an additional day for each span would have increased construction time by almost 50%. Although experience with the production and placement of D3 segments was the base for the agreement with the D6 segments, it clearly meant entering unchartered territory.
- *Organizational decision-making*: All engineers and managers previously involved in the development took part in the final decision meeting. Following an elaborate presentation, the attendants discussed the pros and cons controversially along with the involved risks. In the end, they settled on the adoption of segmental construction. Everything in this solution was (and continues to be) innovative – the substructure, the superstructure, the erection equipment, the transportation equipment, the technology, and the organization.
- *Implementation*: It took the first girder crew only six spans (out of more than 1200) before they achieved the planned erection speed of 2 days for the first time. By then, the project team had overcome most problems with all other innovative features of the product, the processes, and the organization. Improvements were ongoing.

13.2.7 Process Innovations for the Great Belt Tunnel

Awareness of a problem: An important consideration for the site installation of the Great Belt Tunnel was the location of the factory for the tunnel segments. The bid team followed a simple heuristic locating the segment factory close to the tunnel entrance to minimize transports. This was the standard solution for other projects, notably on the French side of the Channel Tunnel.

The site installation team in charge of developing the final layout consisted of young engineers. They had little prior experience in tunnel projects. Therefore, they were unaware of known solutions from previous projects and, hence, could not fall back on the heuristic employed. Their analysis evoked two disturbing facts. All deliveries had to go into the tunnel but a factory close to the tunnel entrance would block this flow, and half of the segments from the factory had to be transported to the other end of the tunnel by ship as it was located on an island and tunnel-boring machines had to be served at both ends.

Analysis of the problem: In order to realize the problem, the site installation team had to advance partly with the analysis of the solution for the bid. The question arose regarding where to best place a segment factory when it had to serve two tunnel entries – one that was directly accessible and one that was only accessible by sea. In addition, it was determined that the most economical way to receive the aggregates for the concrete segments was also by sea. In both instances, it was necessary to construct a temporary harbor at each end of the tunnel (one on the large island of Zealand and one on the small island of Sprogoe). However, there was not enough space for the segment factory at the harbor on Zealand.

Searching the set of known solutions: The known solutions could not provide a satisfactory answer for how to minimize costs and logistics. There was no adequate solution for the problem at hand.

Analysis of the gap: The gap between previous projects and the Great Belt Tunnel was exactly the Great Belt that separated the two tunnel entrances. A factory close to one tunnel entrance would have made it necessary to bring all the aggregates to the factory and take half of the segments back the same way, doubling transportation.

What-if questions: What would happen if the contractor decided to place the factory at the temporary harbor on Zealand? This would allow to deliver incoming aggregates directly to the factory and to transform them into segments there. These could then be distributed to the two tunnel entrances. This solution would minimize transportation lengths and reduce the means of transportation. There would be no need for trucks to transport aggregates from the harbor to the factory at the tunnel entrance. So, if there was no space at the harbor, would it be possible for the contractor to create the space by land reclamation?

Satisficing: Land reclamation would require shipping in sand for an area of 1000 m by 200 m and the construction of a quay for later shipments of aggregates and segments. The site installation team elaborated on this solution and thought it largely preferable to the previous solution. It provided for a separation of the areas of sea transport, prefabrication of segments, and tunnel construction. Having found a solution that was better than the previous, the search stopped. The employed heuristic was to improve the previous solution, not to find the best possible solution. In view of the tight schedule, other issues then needed attention.

Adequacy of resources: The cost of the invention and the feasibility of land reclamation were of concern. In the end, an estimate showed that the solution promised lower costs and the feasibility seemed assured.

Organizational decision-making: By chance, the leader of the site installation team presented the problem and the invention over a beer to two older colleagues who were in charge of the segment factory. Their first reaction was *"This is not how things are done, you don't know because you are too young."* The team leader, however, instead of becoming discouraged, simply explained the pros and cons of his proposition. As the two colleagues could do nothing but listen (and drink their beer), they heard him out. In the course of their discussion, the advantages of the invention became decidedly clear.

Later, an official meeting took place and the young team leader again presented the idea. At first, the top management rejected the idea with the same argument of a lack of knowledge. Only with the help of the two previously briefed experienced engineers was the team leader able to engage his audience long enough to outline the advantages. As they became clear, the group reached a preliminary decision to pursue the idea further. The project management later gave their final approval after checking a detailed report.

Implementation: The implementation posed no problems, although the land reclamation, the construction of the harbor, and the segment factory were impressive temporary works. Other circumstances delayed the start of tunneling, and it proved to be a blessing to have enough space outside the construction area for the production and storage of the segments. This was a possibility, which had been considered in the analysis. What had not been considered was what happened at the end of the tunnel project. At that time, construction works started for the Eastern Bridge of the Great Belt Crossing, and the bridge contractor bought the reclaimed land, the harbor, and the factory building from the tunnel contractor. Otherwise, the tunnel contractor would have been required to dismantle a segment factory at the tunnel entrance at extra cost. Sometimes, there is something like a free lunch. Selling the temporary site installations was an emergent innovation.

13.2.8 Conclusions

The goal was to extract from data a practical model of the innovation process in construction. The model features nine overlapping process groups and the integration of 11 partial plans. This model is richer than those presented in the general and construction literature (Table 13.6). It also contains more qualitative information. The chosen nominal definition of construction innovation covers a process, a focus (product, technology, and technical/management organization), an operationalization (new to the contractor), and an aim (business advantage). The model provides information in all these areas: it describes the process. The focus shifts during the alignment of the 11 partial plans between product, technology, and organization; and, during decision-making, the contractor checks for the business advantage. I do believe the model can help practitioners manage the innovation process much more efficiently. The reason for the usefulness to practitioners is that the model is not normative but descriptive based on data from projects and practitioners.

The descriptive nature becomes clear when looking at the heuristics used (satisficing, expert judgment, and intuition) and biases found such as groupthink or personal interests. The illustrative examples are proof that the practice is not a beacon of ideal rationality but of a rationality determined by circumstances. Time pressures, complexity, and cognitive limitations lead to shortcuts. None of this is new in general as the relevant literature discusses heuristics and biases thoroughly (Kahneman 2011; Thaler 2015). However, it is still helpful to find confirmation in the data for construction megaprojects.

The defining difference between the cases studied by Van de Ven et al. (2008) and my cases is the time pressure under which construction works: contractors must submit bids at a given date and finish the project as stipulated. The time to find a solution is short, and periods of divergent activities exist, but they are sooner than later forced into convergence. Therefore, time places limits on the innovativeness within a project. Contractors compensate this with innovation between projects, from one project to the next, by referring to solutions of the past and building onto them.

The innovation process is complex: the number of possible components, their interrelation, and their cross-impacts are limitless in megaprojects. Only bold decisions can reduce this complexity. The innovation process is iterative: team members try constantly to align the contradictory demands of the 11 partial plans. They must switch back and forth between the different phases in the model. At time t_1 in Figure 13.4, six activities are ongoing simultaneously: problem and gap analyses are not finished, the search for known solutions continues, generation of new solutions and coordination of partial plans has started, and the adequacy of resources requires monitoring. New information in one activity can impact all others. The innovation process is messy: there exist no guidelines on how to proceed. In all cases, the involved managers and engineers had to learn their path through the innovation process anew. Based on observation, there was never enough information, structure, time, or resources – and yet, decisions were due.

Doggedness, gameness, courage, and a certain degree of aggressiveness are characteristics ascribed to pit bull terriers, and these are the qualities required for a successful innovation process in megaprojects. Ingenuity is helpful in the team but not the determining trait.

With regard to national cultures, I found different degrees of willingness to engage in innovation; however, no discrepancies with regard to the process became evident. Procurement methods certainly have an influence on innovativeness but not on the process, as the two examples from the BangNa Expressway (design/build) and the Great Belt Tunnel (design/bid/build) show. The two projects also exhibit different degrees of fragmentation in the supply chain, yet, again, with little impact on the innovation process.

Another important result from the dataset is that contractors always managed innovations through project-based problem solving. I never observed the involvement of an R&D department (I know that they exist in companies, but they had no influence in these cases). People welcomed emergent solutions when they detected them.

We have seen that contractors checked innovative solutions for a business advantage during the decision-making process. The bid or project teams were aware of this hurdle and always considered this aspect before submitting solutions for organizational decision-making. Profit maximization was the guiding principle observed in the innovation processes. We can interpret rational behavior of the actors only in this context: contractors engage in the innovation process for a business advantage. At this point, the doggedness of pit bulls became most obvious.

13.3 Progress Functions

For almost 100 years, researchers and practitioners have analyzed progress curves in diverse manufacturing industries. Unfortunately, no data are available for the complex production processes in the construction industry. The basic concept states that we can reduce production costs by gaining experience. Learning and innovation play an important role. Progress curves are not the result of automatic mechanisms; managers, engineers, and workers must actively pursue them. Production discontinuity is a characteristic that distinguishes the construction industry from manufacturing: it is not evident that the concept of progress curves applies to construction. To expand present knowledge, a cost analysis of four successive expressway projects in Bangkok provides data. This is a single case study design with four embedded units of analysis. The data disclose impressive progress based on cumulative experience. The cost reductions follow almost a 70% slope for every doubling of output. Due to the discreet nature of projects, induced learning by engineers and managers at the beginning of a project grinds out more progress than autonomous learning-by-doing within projects. Although the data reveal a clear trend, they do not permit formulating a mathematical regression function. Progress curves in construction can serve for goal setting, not for predicting future costs. This becomes evident when paying attention to the caveat "it is possible" in the following definition.

Progress curves in construction = (def.) It is possible to reduce the inflation-adjusted cost of construction projects in the magnitude of 20–30% by doubling the accumulated amount produced with one technology.

13.3.1 Theory and Terminology

The concept of progress curves describes the phenomenon that companies can lower production costs as the cumulative amount of goods produced increases. In an early study, Wright (1936) found a cost reduction of 20% when doubling the amount of aircrafts produced; and, in a later study, Couto and Teixeira (2005) discovered a reduction of 15% in buildings.

Progress curves are models of a dynamic production theory (Dutton and Thomas 1984). As such, the theory states that it is possible to reduce the inflation adjusted (real) production costs for a certain product by a fixed percentage α when doubling the cumulative amount produced. An increase in production experience is the driver of the possible cost reductions, and the cumulative production captures the progress of experience. In order to gain experience, the product and the production technology must remain unchanged. The theory also contains a caveat by introducing the words "*it is possible…*" This does not postulate a law. By neglecting this caveat, a simplified equation can represent the theory:

$$c(x) = c_1 (1 - \alpha)^\delta \tag{13.1}$$

In this formula, c_1 denotes the costs of the first unit, α the cost reduction factor, δ the times of doubling output, and $c(x)$ the cost of unit x. If $c_1 = 10$ USD and $\alpha = 0.2$ (a cost reduction of 20%), then unit 1 costs 10 USD (10×0.8^0). When doubling output once, the second unit will cost 8.00 USD (10×0.8^1) when doubling output twice, the fourth unit will cost 6.40 USD (10×0.8^2), etc. The following formula represents the same idea by using cost elasticity b (or a learning factor b):

$$c(x) = c_1 x^{-b} \tag{13.2}$$

In this case, production costs depend on the learning factor b. The learning factor b relates to the cost reduction factor α by

$$-b = \log (1 - \alpha)/\log 2 \tag{13.3}$$

For $\alpha = 0.2$, the learning factor equals $-b = -0.32$ and the cost for unit 1 again 10 USD $(10 \times 1^{-0.32})$, for the second unit 8.00 USD $(10 \times 2^{-0.32})$, and for the fourth unit 6.40 USD $(10 \times 4^{-0.32})$. In sum, a cost reduction of 20% is equivalent to a learning factor $b = 0.32$, and a cost reduction of 30% equivalent to a learning factor $b = 0.51$.

The level to which costs fall each time cumulative output doubles defines the slope of the progress curve. For an 80% progress curve, an increase in production from one to two units means a cost reduction from 100% to 80%; an increase from two units to four means a cost reduction of 80% from the previous level (80%) to 64% of initial costs (Lieberman 1987). Thus, the cost reduction factor α, the learning factor b, and the slope of the progress curve are alternate ways to describe the same phenomenon.

Some confusion prevails with regard to terminology: learning curves, progress curves, and experience curves are three of the terms used to describe the phenomenon. Dutton and Thomas (1984) provide a clarification based on a literature review (Table 13.8). They observe that differences exist regarding the type of improvement (simple or complex change), the unit of analysis (worker, crew, project, firm, or sector), and the value used for comparison (manhour, cost, price, and time).

- Learning curves describe simple changes by single workers or crews measured in manhours or costs.
- Progress curves describe complex changes affecting a project or firm measured in costs.
- Experience curves describe highly complex changes in a sector measured in prices.

Changes on the level of a sector evidently contain changes on levels of worker, crew, project, and company. More simply expressed, learning curves can be effective by themselves, while progress curves always include learning curves. Learning curves reach a normal level after a few repetitions (Uher and Zantis 2011). Progress slows down but never stops. Figure 13.5 shows that progress curves start at much higher costs as they describe complex operations. The cost reduction potential is greater. Finally, the cost reductions

Table 13.8 Differences between learning, progress, and experience curves.

	Learning curve	Progress curve	Experience curve
Type of change	Simple	Complex	Highly complex
Level	Worker or crew	Project or company	Sector
Measurement	Manhours or cost	Cost	Prices

Figure 13.5 Comparison between progress and learning curves.

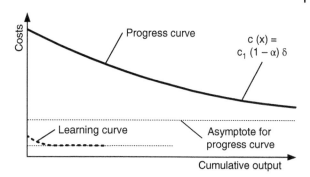

approach an asymptote but never level out. The theory as expressed in Eqs. 13.1 and 13.2 applies only to progress curves (Dutton and Thomas 1984).

The data I will present are from four expressway projects. Therefore, it will be possible to analyze them with regard to learning and progress curves but not with regard to experience curves.

13.3.2 Literature Review

The discussion of progress curves in manufacturing has been ongoing for a long time. The origins go back to the decade between 1920 and 1930. Leading business journals published the intensive research that followed. Two strings are detectable, and they differ with regard to the accuracy of the phenomenon. While practitioners use progress curves to predict future results (thus assuming some relationship close to a natural law), scientific researchers point to contextual influences and a general inaccuracy. They promote using progress curves as a management tool for goal setting. The discussion thinned out after 1990 in manufacturing. Only some authors published further contributions afterwards with findings from specific industries.

As the discussion ebbed away in manufacturing, it surged in construction. In a dedicated literature review on construction, Gottlieb and Haugbølle (2010) identified 27 papers loosely connected with progress curves; 23 of these were published after the year 2000. The discussion in construction as a custom industry must consider a different context and can add to broaden the understanding of the phenomenon. It is not only of interest for the construction industry if it takes account of the stream of arguments elaborated by practitioners and researchers for manufacturing.

Wright studied data on progress curves gathered on airframe production in the 1920s and published the findings in 1936. Alchian (1963) analyzed data from US airframe production during the Second World War. Other industries also attracted the attention of researchers, e.g. machine tools (Hirsch 1952), metal products (Dudley 1972), petrochemicals (Stobaugh and Townsend 1975), chemical processing (Lieberman 1984), shipbuilding (Argote et al. 1990), and photovoltaics (Nemet 2006).

Of the vast number of topics discussed over the past 100 years, five bear specific importance for construction research: (i) the slope of the progress curve, (ii) the form of the progress curve, (iii) possible reason for cost reductions, (iv) types of learning contributing to

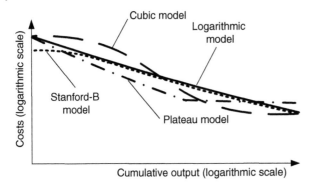

Figure 13.6 Different types of progress curves discussed in the literature.

cost reductions, and (v) differences between manufacturing and construction with regard to progress curves.

The slope of the progress curve can be appreciated by looking at some samples: Ford Model T (86%); aircraft assembly (80%); steel production (79%); equipment maintenance (76%); hand-held calculators (74%); and integrated circuits (72%) (Cunningham 1980). Gottlieb and Haugbølle (2010) summarize the findings from buildings at a level of 93–87%.

The form of the progress curve can differ as researchers have proposed four models (Yelle 1979). The logarithmic model is the standard introduced above. Figure 13.6 depicts it in a double logarithmic graph where it becomes a straight line (Wright 1936). Carr (1946) proposed a cubic model based on production by crews with different levels of experience. Garg and Miliman (1961) introduced the Stanford-B model based on the production of Boeings 707. Baloff (1971) found a plateau model to be representative of labor-intensive industries.

The reasons often cited for cost reductions are learning, innovation, and economies of scale (Dutton and Thomas 1984). Learning and innovating strongly influence each other. I have already explained that I see learning – also with regard to management – as process and innovation as result.

Economies of scale are an alternative explanation for cost reductions. They depend on current output rather than on cumulative output (Hall and Howell 1985). One must know the production volume ex-ante to realize economies of scale (Dutton and Thomas 1984) because the decision for the production system precedes production, and the minimum efficient scale depends on throughput. Hollander (1965) found that only 10–15% of cost reductions in a chemical plant were due to economies of scale. Day and Montgomery confirm these data (1983).

Types of learning contributing to cost reductions are more placeholders than thorough descriptions of learning processes. Cost data used for the analysis of progress curves do not contain information about learning. Dutton and Thomas (1984; cf. Section 10.4) distinguish between the source of learning (exogenous or endogenous with regard to the project or the firm) and the way of learning (autonomous or induced). Autonomous learning happens within an existing set of rules and induced learning by also changing these rules.

Differences between manufacturing and construction with regard to progress curves can build on insights from manufacturing. For the latter, continuous production processes within one factory provide the data on the slope of progress curves. When producing different (but similar) products in different places, Alchian (1963) found different progress curves. He speaks of specific airplane/factory combinations. This resembles construction, where we build different structures in different places. In consequence, we would expect specific structure/site combinations. The learning curve (i.e. few repetitions and plateau) is well-known in construction and explained in textbooks for the purposes of cost control (Barrie and Paulson 1992) or scheduling (Uher and Zantis 2011). It is especially pertinent to line-of-balance scheduling (Arditi et al. 2002). The same is not true for progress curves. Only one recent study by Couto and Teixeira (2005) provides primary data when investigating two housing projects with 8 and 12 floors.

13.3.3 Research

If progress curves are widespread phenomena in continuous production processes and if there are separate progress curves for specific product/factory combinations, the question is what happens between these extremes. This question is pertinent to construction. We can expect progress curves within one project, which is a specific structure/site combination. Can we also expect progress curves continuing from project to project where there are at least some shared resources? These thoughts lead to three questions:

- Can we find cost reductions within consecutive construction projects that are a consequence of learning?
- If such cost reductions exist, what is the slope of the progress curve over consecutive projects?
- If such a progress curve exists, what is its shape?

Until today, there has been no published data allowing an answer to any of these questions.

The hypothesis formulating the progress curve theory includes possibility as a caveat (it is possible to reduce costs…). A project team might or might not achieve cost reductions according to this hypothesis. Given such a caveat, rigorous falsification becomes impossible. Popper (2002) postulates falsification as selection criterion for a scientific approach. To him, the theory of progress curves lies clearly outside the realm of science. However, one can argue with Lakatos (1970) that this caveat does not touch the hard core of the theory (cost reduction due to production experience) and that plausible explanations of deviations in the dataset from the ideal progress curve can serve as an auxiliary hypothesis. It then depends all on the plausibility of the argument.

When measuring cost reductions, one must analyze quantitative costs. Such data seem to be reliable and relatively easy to analyze. However, they are more complex than meets the eye and require interpretation. The route from the actual event of pouring concrete on site to reflecting this activity in cost data is lengthy and difficult. The translation of an activity into monetary values requires numerous assumptions, abbreviations, and modeling; it is never a one-to-one relationship. In addition, even straightforward cost analyses are part of company politics. Therefore, it becomes necessary to completely understand the significance of every cost item. In sum, every researcher of progress curves invariably faces two problems: (i) a caveat in the formulation of the hypothesis, and (ii) a problem of cost interpretation.

I will present and analyze the data of four consecutive expressway projects in Bangkok, Thailand. The contractor assembled these data for purposes of cost control. As such, they allow for a quantitative progress curve analysis. They do not contain background information. In accordance with the research questions and the problem of fuzzy data, I conducted clarification interviews to make sure that I understand the provenance and meaning of the data completely. Interview partners were those with the best knowledge for a specific question, and these were project directors from all four projects, project team members, the cost controller responsible for the inter-project comparison, as well as the contractors' top and middle management.

The projects started in the decade between 1990 and 2000. The boundary conditions for all projects were similar: the same owner contracted the same joint venture to design/build contracts. The location, climate, economy, and resources were also the same. The construction technology – segmental bridge construction – was comparable.

A strong ceteris paribus clause applies to the four embedded cases (owner, contractor, procurement method, contracts, product, technology, location, climate, economy, and resources). The four projects (A, B, C, and D) were all megaprojects with a large degree of complexity; the project values ranged from 218 to 1000 million USD. They commenced in a chronological sequence with one break between A/B and overlaps between B, C, and D (Figure 13.7). The commonalities permit searching for repeated evidence by literal replication. Despite all commonalities, project C stands out by noticeable differences in size, technology, organization, and people. This provides a chance for theoretical replication between projects A, B, and D and project C. Table 13.9 gives an overview of the research possibilities.

For project A, only general data exist. It is not advisable to present absolute monetary values as cost data are proprietary. Relative values can serve the same research purpose when looking for the magnitude of the slope of the progress curve and causal relationships.

It is rare in construction that we can compare any two projects directly. This statement also holds true for the four expressway projects despite the ceteris paribus clause. It is a painstaking part of research on the progress curve to identify differences and make the necessary adjustments in order to avoid comparing apples to pears. It is also indispensable. The presentation of the raw quantitative data and the adjustments are the next topic. The conducted interviews provided much of the information on the adjustments. While this

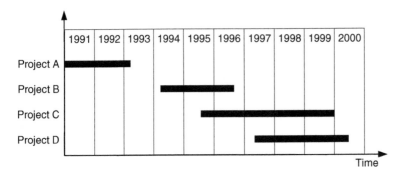

Figure 13.7 Schedule for the four expressway projects.

Table 13.9 Research possibilities for the four expressway projects.

	Project A	Project B	Project D	Project C
Project size (m² bridge deck)	220	365	248	1900
Technology, organization, people	Similar between A, B, and D			Different
Literal replication	Between A, B, and D			Not possible
Theoretical replication	Between A, B, and D and C			
Question 1 (cost reductions in projects)	Compare all projects			
Question 2 (slope of progress curve)	Combine all projects			
Question 3 (shape of progress curve)	Combine all projects			

analysis serves to format the raw data, the following one will build on the refined data for analysis.

A comparison between the four projects yields the following data: (i) general data showing the commonalities and differences of the projects, (ii) time sequence of the projects, (iii) quantity differences indicating project differences or design progress, (iv) technological differences specifying technology advances, and (v) cost data.

Table 13.10 provides general data to show the commonalities and differences of the projects. Commonalities not shown are the geography and climate (Bangkok), the economy (Thailand), the owner (ETA), the contractors (German–Thai Joint Venture), the contract form (design/build), the type of structure (expressway), and the overall technology (segmental bridge construction). Some differences between the projects provide insight, whereas others blur the picture. The former group comprises variances in size, complexity, and contract price; the latter includes variances in bridge height, in-situ soil conditions, local traffic and utility relocation, and interference with other construction works.

Table 13.10 Characteristics of the four expressway projects.

	Project A	Project B	Project C	Project D
Location	Bangkok, Thailand			
Owner	Rapid Transit and Expressway Authority of Thailand (ETA)			
Contractors	German–Thai Joint Venture			
Contract form	FIDIC Design/build Contract			
Contract volume (million US$)	218	383	1000	244
Superstructure length (m)	20	35	55	10
Bridge deck area (thousand m²)	220	365	1900	248
Average width (m)	11	10	35	24,8
Comparative real price (US$ m⁻²)	1	890	433	737
Average height (m)	12	12	18	12

Table 13.11 Quantity comparison.

	Project A/B	Project C	Project D
Concrete volume substructure ($m^3\,m^{-2}$)	0.46	0.46	0.37
Reinforcement substructure ($kg\,m^{-2}$)	85.0	53.0	44.0
Concrete volume superstructure ($m^3\,m^{-2}$)	0.51	0.49	0.51
Reinforcement superstructure ($kg\,m^{-2}$)	71.0	59.0	71.0
Post-tensioning superstructure ($kg\,m^{-2}$)	24.4	24.0	24.4

The time sequence of the projects starts in 1990 and ends in 2000. The 14 months between the end of project A and the beginning of B permits the study of the impact of a production gap. As the other three projects overlapped and as people with their knowledge switched between them, experience transfers from project to project became possible (Figure 13.7).

Quantity differences indicate differences between projects or design innovation. Primarily, the data in Table 13.11 show differences between project C and the others. These are due to design changes, as can be seen in Figure 13.8 by comparing the left (projects A, B, and D) and the right picture (project C). Differences also exist between projects A and B on the one side and D on the other with regard to the substructure. This is a consequence of project specifics; projects A and B have an average width of around 10 m and project D of 25 m. A smaller expressway requires more columns and foundations; thus, substructure quantities are higher for projects A and B. Project C requires the same amount of concrete and less reinforcement than projects A and B but for the substantially greater bridge height of 18 m instead of 12 m.

While the overall technology was the same for all projects, there were some noticeable variations with regard to substructure, superstructure, and erection equipment between the

Figure 13.8 Comparison D3 overhead girder (left) and D& underslung girder (right).

projects (Figure 13.8). The left and the right pictures clearly demonstrate the differences. In addition to the different shape of the columns, one needs to keep in mind the greater height in project C. The segments for projects A, B, and C have a maximum width of 15.30 m for three lanes (D3); the segments for project C are 27.20 m wide for six lanes (D6). The contractor exclusively used overhead girders for projects A, B, and D; for project C, it was a mix of six overhead girders (600 000 m²) and five underslung girders (1 300 000 m²). Project C is also the largest one with regard to the use of D3 overhead girders (600 000 m² vs. 383 000 m² for project B). In the dataset, projects A, B, and D form a core family; project C is a relative.

There are numerous problems with data for the cost evaluation of progress curves in construction: (i) production unit, (ii) costs, (iii) construction technology, (iv) starting point of the progress curve, and (v) construction inflation.

Products in construction are large, expensive, and unique. We cannot easily compare between megaprojects. Product size and singularity are important differences to mass production. In quantity surveying, the usual yardstick to compare costs are square meters or cubic meters. Accordingly, the production unit of choice for bridges is a square meter of superstructure – or deck – area and not the project itself. When comparing the deck areas of bridges, three further aspects must be considered that impact the cost per square meter. The span length is of utmost importance as bending moments increase with the square of the length. Soil conditions can almost be as important as span length. The bridge height is of comparatively lesser significance. At around 42 m, the span length is similar for all four projects and soil conditions are also comparable. The greater bridge height of project C necessitates an adjustment of the cost per square meter.

We need cost data from the contractor as producer to build a progress curve. The difference between the owner's contract price and the contractor's cost is the contractor's profit or loss. Project A had a neutral result, project B, a profit of 19%, project C, a loss of 34%, and project D, a profit of 3%. It becomes evident that working with prices (owner's costs) distorts the data beyond repair. For the four chosen projects, cost data are available. Utility relocation, taxes and fees, interest during construction, and company overhead costs are not directly related to production and will be subtracted from total costs for the purpose of comparison. The precast yard for the projects A, B, and D was provided by the owner; therefore, the costs for the precast yard for Project C are also subtracted. An analysis of the costs for the additional height of project C shows that these amount to 14% of the substructure costs, and this is subtracted from the project costs. Detailed cost data for Project A are not available. As the total costs for project B were 19% less than for project A, we will assume that this applies to all items. Further adjustment to the cost data is necessary to account for claims and construction speed.

Strictly speaking, we can never employ the same technology in construction. The site layout and, with it, the site installation (as factory) as well as the soil conditions always depend on the project. What constitutes a technology needs to be defined. In the dataset, we find projects A, B, and D using overhead girders and project C with a combination of underslung and overhead girders. The data must decide to what degree this constitutes comparable technologies.

In a mature industry such as construction, a contractor will possibly have some earlier experience with a technology. This makes it difficult to determine the starting point of the progress curve and has an impact on the run of the curve. According to interviews, it seems

Table 13.12 Comparative costs of the four expressway projects.

	Project A (D3)	Project B (D3)	Project D (D3)	Project C (D3/D6)	Range
Total costs	100%	65%	50%	38%	62%
Plant, equipment, and site installation	100%	42%	11%	4%	96%
Design costs	100%	68%	74%	25%	75%
Management costs	100%	68%	39%	29%	71%
Material costs	100%	68%	59%	45%	55%
Labor costs	100%	68%	59%	52%	48%
Financial costs	100%	68%	46%	78%	22%

reasonable to assume a remembered production experience of $100\,000\ m^2$ by the contractor and to choose this as starting point.

General inflation data are easy to obtain; however, often, the construction business cycle is not aligned with the average consumer price index (CPI). In addition, housing, industrial building, and civil engineering follow their own price development paths, where considerable differences exist. Due to a lack of more detailed data, I will make use of CPI inflation data from the World Bank to adjust the costs (World Bank 2015). The start of a project is the point where inflation counts because the estimated costs include price increases during construction as escalation. The inflation factors are 1.00 for project A, 1.17 for project B, 1.25 for project C, and 1.34 for project D. Table 13.12 lists the real costs as percentages for different items considering technical adjustments.

Table 13.12 shows projects A, B, and D in sequence because they use a similar construction technology. Data for project C appear in the following column. The range indicates the magnitude of improvements for the different cost items relative to project A.

13.3.4 Data Analysis and Discussion

The data analysis focuses on three possible contributors to the cost reductions of Table 13.12: economies of scale, innovation, and learning.

Economies of scale are difficult to separate from cost data (Day and Montgomery 1983). From the business literature, we know that they contribute comparatively little to the reductions. Nevertheless, they pose the problem of an alternative explanation for a part of cost reductions. The question with economies of scale is what happens when production is pushed from level $x^0 = f\,(r_1{}^0 \cdots r_i{}^0)$ to level $x^1 = f\,(\lambda r_1{}^0 \cdots \lambda r_i{}^0)$, with r being production factors and λ a proportionality factor. If the output grows more than the inputs (with a higher factor than λ), there are economies of scale; if it grows less, there are diseconomies of scale. In the case of economies of scale, fixed costs for plant and equipment should become a smaller percentage of total costs. Design and management costs might also add to economies of scale. The data in Table 13.13 are inconclusive.

Project B shows indications of diseconomies of scale compared with project D, while a comparison between projects B and C might point to economies of scale. Engineers plan

Table 13.13 Selected costs in percent of total project costs by size.

	Size (m²)	Fixed costs (%)	Management costs (%)	Design costs (%)
Project A	220 000	13	15	3
Project D	248 000	3	11	4
Project B	365 000	8	15	3
Project C	1 900 000	1	10	1

construction projects for optimum size, and there should therefore be continuous results. The up and down of the values in Table 13.13 find no consistent explanation based on scale effects. On the other hand, the results are perfectly in order with learning effects. When arranging the fixed costs chronologically, we get 13% (A) – 8% (B) – 3% (D) – 1% (C).

Innovation is clearly visible in projects A, B, and D as compared to previous ones. Again, project C was much more innovative in contrast to the other three. Most innovations were the result of induced learning. Although the contractor employed design consultants and a few subcontractors, he always determined the outcome. For all projects, contractors buy materials and equipment; they will buy whatever is best suited for the project. Accordingly, among the purchases can be external (off the shelf) innovations. In the projects considered, these had no noticeable influence on costs. External innovations played a small role in reducing the costs.

Learning is clearly visible from the data. Ordering the data from Table 13.12 according to the magnitude of the improvement ranges provides an idea of the cost items most susceptible to reductions: (i) plant, equipment, site installation – 96%; (ii) design – 75%; (iii) management – 71%; (iv) material – 55%; (v) labor – 48%; and (vi) financing – 22%.

More obscure is the way of learning (autonomous or induced). All projects were design/build contracts. The contractor and the designer developed the design jointly, and the contractor provided production expertise to improve buildability. Projects A, B, and D used similar designs, equipment and plant, processes, technologies, and organization. While the top management changed between project A and B, the same group moved from project B to D, together with most of the labor crews. Project C was quite different from the others with regard to design, technology, and organization. All information from the interviews suggest a learning path from project A to B to D. Project C commenced based on existing knowledge at the time of its inception learned from projects A and B. However, the data show that project D did not benefit in any way from the preceding project C.

As construction is all about discontinuous production due to the project organization of the industry, the easiest way to induce learning is when setting up a new project. Induced learning can also emerge during construction; autonomous learning always happens during construction.

All projects show numerous learning curves leveling out in a plateau after a limited number of repetitions. The impact of such learning curves is too small to detect in the cost data, but they are discernible in schedules and schedule control. Project planning considered such effects for all starts of launching girders. The most telling example is the first use of a D6 underslung girder. After the learning period, the schedule planned with a cycle time of

Table 13.14 Data for the evaluation of different ways of learning.

	Start	Induced learning	Autonomous learning
Project A beginning	1990	0	???
Project A end	1993		
Project B beginning	1994	28%	7%
Project B end	1996		
Project C beginning	1995	22%	8%
Project C beginning	2000		
Project D beginning	1997	10%	5%
Project D end	2000		

2 days per span. The actual sequence was as follows: 36/6/4/3/4/2 days. After 10 cycles, the effect leveled off to a constant 2-day cycle. The 36 days for the first span are not indicative of erection works alone; they included completion of the underslung girder.

Learning was exclusively a result of endogenous activities, both autonomous and induced (Table 13.14). The data show the difference between the previous project and the focal one. The data for B indicate the difference to A and the data for D, the difference to B as these are in a chronological order on a learning thread. Project C capitalizes upon the experience available at the time of its inception, i.e. a midway point for project B.

Autonomous learning is learning-by-doing; this is relevant in labor and management costs in Table 13.12. The quantity and quality of material costs depend on the design, while the price depends on the market. At the beginning of a project, managers and engineers decide on plant, equipment, site installation, design, material, and financing. These were internal decisions in all four projects. Thus, all ensuing improvements are due to endogenous induced learning.

The data in Table 13.14 allow the construction of the progress curve for the four expressway projects (Figure 13.9).

13.3.5 Discussion and Conclusion

Learning is obvious and considerable; most points of the progress curve are below the line depicting a 70% slope. Furthermore, progress in the four observed projects is considerably greater than what previous authors have reported from their research in the construction industry. The cost reduction was 35% for project B, 30% for project C, and 15% for project D. Another apparent point is that induced learning at the beginning of the project is far more important than autonomous learning during the project: in the beginning, managers and engineers can influence progress substantially.

We can distinguish two progress curves, one from project A to B to D and a second one from project A to B to C. Project D did not benefit from lessons learned in project C. This is clearly an opportunity missed. From interviews, we know that noteworthy spillovers did not occur between project C and D.

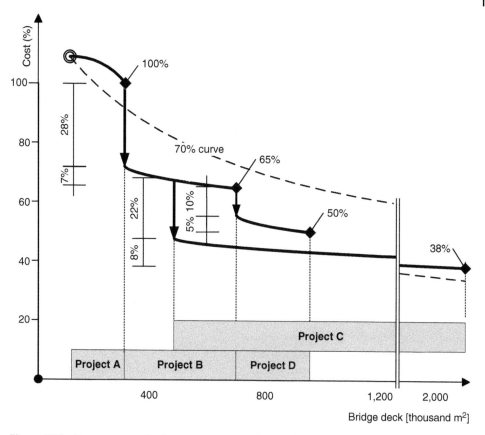

Figure 13.9 Progress curve for four expressway projects in Bangkok, Thailand.

The crew supervisors for each girder proved to be the core knowledge group for the superstructure erection. Although only 25% of the supervisors in project C had prior experience with construction technology, the crews reached the cycle of 2 days per span for the first time after six spans and consistently after 10. Given that this girder erected 310 spans in total, this learning curve hardly matters. In retrospect, the two technologies seem similar. There appears to be a certain robustness as to what constitutes a technology.

No detailed quantitative data are available for project A. Qualitative data suggest an inverse progress curve. There were considerable problems to achieve the desired construction speed for the superstructure with a 2-day cycle. It also took a long time to master the erection technology. In such a complex production process, lack of experience impeded improvements. This is in accordance with the assumptions for a cubic and the Stanford-B model (Figure 13.9).

The contractor was not aware of progress curve effects in the decision-making process of all four projects. Project A was not a success; profits were zero. This put a lot of pressure on the preparation of project B. By then, mastering the erection technology no longer posed problems. However, in order to make the project financially successful, more changes were necessary. The effort provided for a cost reduction of 28% due to induced and 7% due to autonomous learning. Project B became a great success, and the management team received substantial praise. The reduction efforts clearly outpaced the reduction in market price. The contractor judged the market situation for project C to be different compared to the previous projects and decided on predatory pricing. The submitted price in real terms was not even half the price for project B, which started only 15 months earlier. This again led to considerable efforts in cost reduction. The project incurred a sizeable loss and pressure to reduce costs further while the execution never ceased. Then, instead of capitalizing on the experience gained through project C, project D was a continuation of project B; the successful team just moved on, transferring the organization, technology, and design concepts; the profit dropped significantly. Although the consideration of the progress curve downgrades project D to a failure, the contractor considered it a success because it still generated a profit. Perceived market prices drove the strategy; the limits imposed by a progress curve played no role, nor did the opportunities.

If all projects have the same learning rate, they would closely follow the 70% slope in our sample. They do not. The end of two projects are above this line and two are below. The learning rate differs. Therefore, it is not advisable to use a progress curve to predict costs at the end of a project; the variations are considerable. On the other hand, the progress curve can serve as a guideline for internal cost management by setting a goal. If this goal cannot be achieved, an explanation is required. This warning is in accordance with the academic literature, but not with publications by practitioners.

The presented results allow construction to follow the same strategies as manufacturing if a company is able to procure sequential work for one type of structure or one technology. The strategies are cost leadership and market reorganization toward an oligopoly. However, there is neither hope nor fear warranted; large volumes of repetitive work are not available in construction. What contractors can achieve is a short-term cost advantage providing higher profits or higher competitiveness. What clients can realize are lower costs through sequential contracting.

14

All in All, What Does It Mean?

One of my favorite American authors is Philip Roth, and in his book *American Pastoral* (p. 63), he writes, "Writing turns you into somebody who's always wrong. The illusion that you may get it right someday is the perversity that draws you on. What else could? As pathological phenomena go, it doesn't completely wreck your life." I feel it is worthwhile to think about this comment, but also that Roth is a bit too critical. Certainly, the wish to "get it right someday" has pulled me along for more than 30 years, from one megaproject to the next. What I want to get right is construction project management for megaprojects. This is the task of contractors. If thinking and writing about a problem with the wish to "get it right" is indeed a pathological phenomenon, I must say that I enjoy this quite a bit. However, this is only one side of it. Thousands of managers and engineers represent the other side: they want to "get it right" predominantly by doing. With a bit of cynicism, one might call this wish to do a megaproject "right" an elusive endeavor, another pathological phenomenon. However, I like to encourage these managers and engineers to keep pursuing their goals with determination and joy while learning from the inevitable mistakes. We must accept that we can never manage megaprojects right. Getting closer and closer to the ideal is already rewarding.

Agrawal (2018) calls her participation in the Shard project in London a *"chance of a lifetime."* I agree with her completely; participation in megaprojects is a chance of a lifetime. Accordingly, involvement in more than one megaproject means even more.

The beginning of all megaprojects is chaos, a complete lack of order. This is true for the front-end (owner) as well as the back-end (contractor). Both owners and contractors have to shape the megaproject within their respective frames. The owner faces a regulatory frame provided by society; the contractor must find solutions within the same regulatory frame, plus the additional constraints imposed by the owner. The owner reduces task complexity to some degree; the contractor faces reduced complexity, but drives it toward zero by placing a physical object in the world. Both create order from chaos, and this is an overwhelming experience. The owner predominantly manages information, while the contractor manages production. Whoever thinks of *The Book of Genesis* in this context is close to understanding the joys of megaproject management. This comes as close as humans can approach the experience of genesis. Unfortunately, we cannot rest after 6 days, and the result will always be imperfect.

Advanced Construction Project Management: The Complexity of Megaprojects,
First Edition. Christian Brockmann.
© 2021 John Wiley & Sons Ltd. Published 2021 by John Wiley & Sons Ltd.

I know that I am quite emotional at this moment. It serves to make a point: the paradigm of unlimited rationality, the model of the rational actor as employed in traditional economics, is a guideline for failure in megaprojects. Instead, they require us to give all that we can possibly provide as input: dreams as well as bone-dry data, analysis as well as intuition, rationality as well as feeling, the use of all our senses as well as all our cognitive capacities, day and night. Megaprojects demand complete immersion. No one in the top management team can close the office door and switch to family mode. We must completely integrate private and business life. This does not mean that either life must suffer on behalf of the other. At best, it means that the private and the business spheres enhance each other.

Many megaprojects push the involved managers and engineers to the edge of their abilities. Consequently, they make mistakes quite often. It becomes important to learn from these mistakes and move on, ready to learn from the next mistakes. While mistakes are acceptable, repeating the same mistakes is not. This requires humility and resilience from the managers. Some of them are world champions with mentions in *The Guinness Book of World Records*. This does not make humility an easy trait to carry.

If it is true that a purely rational approach to managing megaprojects is a precursor to failure, then we must also abandon the predominance of rationality in describing and understanding megaprojects. Of course, rationality is the paradigm of the scientific approach. Consequently, science is not good enough for megaprojects. We need myth and logic, stories and data. To the academic, stories are enlightening as they know their data; to the practitioner, stories *are* data. An example is the telling discovery of "hidden innovations" by Abbott et al. (2007). To practitioners, such innovations were always in the open. The methodology that academics used for finding innovations earlier, the net they cast to capture them, was not appropriate; the mesh proved to be too wide. The practitioners face another problem: they often do not have a clear understanding of the term "innovation." They might only connect something world-changing such as penicillin with the idea of innovation. Comparing the improvements to a construction technology with the innovation of "penicillin" does not allow the practitioners to understand technology change as innovation. The academic net and the practical interpretation need to be aligned before the phenomenon of innovation can be properly understood.

Managers in megaprojects face extraordinary time pressures; they find little spare time to reflect on their experiences. However, they organize their experiences in cognitive maps that they use for orientation and decision-making. These cognitive maps contain implicit knowledge, i.e. hidden information. I understand the cognitive maps as our treasure boxes, and we need a key to open the boxes. The stories (myths) told by megaproject practitioners are this key. Once we have access to the treasure box through stories, we need to evaluate the contents. Here, analysis (logic) – the domain of science – is helpful. To understand megaprojects, we need cooperation between practitioners and academics.

References

Abbott, C., Barrett, P., Ruddock, L. and Sexton M. (2007): Hidden Innovation in the Construction and Property Sectors. RICS Research Paper Series (7/20).

Acemoglu, D. and Robinson, J. (2012). *Why Nations Fail*. New York: Crown Business.

Agrawal, R. (2018). *Built: The Hidden Stories Behind Our Structures*. London: Bloomsbury.

Alchian, A. (1963). Reliability of progress curves in airframe production. *Econometrica* 31: 679–693.

Antoniadis, D., Edum-Fotwe, F., and Thorpe, A. (2012). Structuring of project teams and complexity. *Project Perspectives* 2012: 78–85.

Arditi, D., Tokdemir, O., and Suh, K. (2002). Challenges in line-of balance scheduling. *Journal of Construction Engineering and Management* 128: 545–556.

Argote, L., Beckman, S., and Epple, D. (1990). The persistence and transfer of learning in industrial settings. *Management Science* 36: 140–154.

Argyris, L. and Schön, D. (1978). *Organisational Learning: A Theory of Action Perspective*. Reading: Addison-Wesley Longman.

Baccarini, D. (1996). The concept of project complexity – a review. *International Journal of Project Management* 14 (4): 201–204.

Bacon, F. (1620, 2000). *The New Organon*. Cambridge: Cambridge University Press.

Badger, W. and Mulligan, D. (1995). Rationale and benefits associated with international alliances. *Journal of Construction Engineering and Management* 121 (1): 100–111.

Baloff, N. (1971). Extension of the learning curve: some empirical results. *Operational Research Quarterly* 22: 329–340.

Baregheh, A., Rowley, J., and Sambrook, S. (2009). Towards a multi-disciplinary definition of innovation. *Management Decision* 47 (8): 1323–1339.

Barrie, D. and Paulson, B. (1992). *Professional Construction Management*. New York: McGraw-Hill.

Bateman, T. and Snell, S. (2002). *Management: Competing in the New Era*. Boston: McGraw Hill Irwin.

Bennett, J. (1991). *International Construction Project Management: General Theory and Practice*. Oxford: Butterworth-Heinemann.

Berger, P. and Luckmann, T. (1967). *The Social Construction of Reality: A Treatise in the Sociology of Knowledge*. Garden City: Anchor Books.

Advanced Construction Project Management: The Complexity of Megaprojects,
First Edition. Christian Brockmann.
© 2021 John Wiley & Sons Ltd. Published 2021 by John Wiley & Sons Ltd.

Berkeley, G. (1710). *A Treatise Concerning the Principles of Human Knowledge*. Three Dialogues between Hylas and Philonous. Chicago: Open Court Publishing, 1986.

Bezos, J. (2016): 2016 Letter to Shareholders. Online accessed January 2020: https://blog. http://aboutamazon.com/company-news/2016-letter-to-shareholders.

Blake, R. and Mouton, J. (1964). *The Managerial Grid: The Key to Leadership Excellence*. Houston: Gulf Publishing.

Blayse, A. and Manley, K. (2004). Key influences on construction innovation. *Construction Innovation* 4 (3): 143–154.

Bröchner, J. (2010). Construction contractors as service innovators. *Building Research and Information* 38 (3): 235–246.

Brockmann, C. (2000): Precast elements for segmental bridge construction of the Bang Na Expressway, Bangkok. Proceedings, PCI/FHWA/FIB International Symposium on High Performance Concrete, Orlando, 368–376.

Brockmann, C. (2001). Temperature-induced and time-dependent deformations in wide-spanning bridge segments. In: *Proceedings, Vth Congress of the Croatian Structural Engineers*, 79–84. Brijuni.

Brockmann, C. (2010). Sensemaking in international construction joint ventures. In: *Proceedings of the 18th CIB World Building Congress*, 252–263. Rotterdam: In-house Publishing.

Brockmann, C. (2013): Requirements for Firms Designing Megaprojects. Proceedings of the 2013 CIB World Congress, Paper 46.

Brockmann, C. and Girmscheid, G. (2008). The inherent complexity of large scale engineering projects. *Project Perspectives* 2008: 22–26.

Brockmann, C. and Kähkönen, K. (2012). Evaluating construction project complexity. In: *Proceedings, MCRP Congress 2012*, 716–727. Montréal.

Brockmann, C. and Rogenhofer, H. (2000). Bang Na Expressway, Bangkok, Thailand – World's longest bridge and largest precasting operation. *PCI Journal* 45 (1): 26–38.

Brockmann, C., Brezinski, H., and Erbe, A. (2016). Innovation in construction megaprojects. *ASCE Journal of Construction Engineering and Management* https://doi.org/10.1061/ (ASCE)CO.1943-7862.0001168.

Bruzelius, N., Flyvbjerg, B., and Rothengatter, W. (2002). Big decisions, big risks. Improving accountability in mega projects. *Transport Policy* 9 (2): 143–154.

Büchel, B., Prange, C., Probst, G., and Rühling, C. (1998). *International Joint Venture Management: Learning to Cooperate and Cooperating to Learn*. Singapore: Wiley.

Carr, G. (1946). Peacetime cost estimating requires new learning curves. *Aviation* 45 (4): 76–77.

Chinyio, E. and Olomolaiye, P. (2010). *Construction Stakeholder Management*. Chichester: Wiley-Blackwell.

Coughlan, P. and Coghlan, D. (2002). Action research for operations management. *International Journal of Operations and Production Management* 22 (2): 220–240.

Couto, J. and Teixeira, J. (2005). Using linear model for learning curve effect on high-rise floor construction. *Construction Management and Economics* 23: 355–364.

Cunningham, J. (1980). Using the learning curve as a management tool. *IEEE Spectrum* 17 (6): 45–48.

Cyert, R. and March, J. (1992). *A Behavioral Theory of the Firm*. Malden: Blackwell.

Dainty, A., Moore, D., and Murray, M. (2006). *Communication in Construction*. Abingdon: Routledge.

Day, G. and Montgomery, D. (1983). Diagnosing the experience curve. *Journal of Marketing* 47: 44–58.

Desjardin, J. (2017): 9 of the World's Largest Megaprojects That Are Under Construction. (https://www.businessinsider.de/the-worlds-largest-megaprojects-2017-1?r=US&IR=T).

Dibner, D. and Lemer, A. (1992). *The Role of Public Agencies in Fostering New Technology and Innovation in Building*. Washington DC, USA: National Academy of Sciences Press.

Doty, D. and Glick, W. (1994). Typologies as a unique form of theory building: toward improved understanding and modelling. *Academy of Management Review* 19 (2): 230–251.

Dubois, A. and Gadde, L. (2002). The construction industry as a loosely coupled system: implications for productivity and innovation. *Construction Management and Economics* 20 (7): 621–632.

Dudley, L. (1972). Learning and productivity changes in metal products. *American Economic Review* 62: 662–669.

Dutton, J. and Thomas, A. (1984). Treating progress functions as a managerial opportunity. *Academy of Management Review* 9 (2): 235–247.

Eisele, J. (1995). *Erfolgsfaktoren des Joint-Venture-Managements*. Wiesbaden: Gabler.

Erichson, B. and Hammann, P. (2005). Beschaffung und Aufbereitung von Informationen. In: *Allgemeine Betriebswirtschaftslehre, Band 2* (eds. F. Bea, B. Friedl and M. Schweitzer), 337–393. Stuttgart: Lucius and Lucius.

Fiedler, F. (1964). A contingency model of leadership effectiveness. *Advances in Experimental Social Psychology* 1: 149–190.

Fischer, O., and Krill, A. (1998): The BangNa Expressway – A Full-scale Loading Test of a Precast Segmental Box Girder Bridge for 6 Lanes of Traffic." Proceedings, XIIIth FIP-Congress, Amsterdam, 503–506.

Flyvbjerg, B. (2014). What you should know about megaprojects and why: an overview. *Project Management Journal* 45 (2): 6–19.

Flyvbjerg, B., Bruzelius, N., and Rothengatter, W. (2003). *Megaprojects and Risks: An Anatomy of Ambition*. Cambridge: Cambridge University Press.

Freeman, C. and Soete, L. (1997). *The Economics of Industrial Innovation*, 3e. London: Continuum.

Gambatese, J. and Hallowell, M. (2011). Enabling and measuring innovation in the construction industry. *Construction Management and Economics* 29 (6): 553–567.

Gamson, W., Croteau, D., Hoynes, W. and Sasson, T. (1992). Media images and the social construction of reality. *Annual Review of Sociology* 18: 373–39.

Gann, D. (2003). Guest editorial: innovation in the built environment. *Construction Management and Economics* 21 (6): 553–555.

Gann, D. and Salter, A. (2000). Innovation in project-based, service enhanced firms: the construction of complex products and systems. *Research Policy* 29 (7): 955–972.

Garg, A. and Miliman, P. (1961). The aircraft progress curve modified for design changes. *Journal of Industrial Engineering* 12: 23–27.

Geertz, C. (1973). *The Interpretation of Cultures*. New York: Basic Books.

Geraldi, J. (2008). The thermometer of complexity. *Project Perspectives* 2008: 4–9.

Gibbons, R. (1992). *Game Theory for Applied Economists*. Princeton: Princeton University Press.

Gidado, K. (1996). Project complexity: the focal point of construction production planning. *Construction Management and Economics* 14 (3): 213–225.

Girmscheid, G. and Brockmann, C. (2010). Inter- and intraorganizational trust in international construction joint ventures. *ASCE Journal of Construction Engineering and Management* 136 (3): 353–360.

Gottlieb, S., and Haugbølle, K. (2010): *The Repetition Effect in Building and Construction Works*. Hørsholm: Danish Building.

Greiman, V. (2013). *Megaproject Management: Lessons on Risk and Project Management from the Big Dig*. Hoboken: Wiley.

Grossman, S. and Hart, O. (1986). The costs and benefits of ownership: a theory of vertical and lateral integration. *Journal of Political Economy* 94 (4): 691–719.

Grün, O. (2004a). Das Management von Großprojekten: Der Widerspenstigen Zähmung. *Zeitschrift Führung + Organisation* 73 (6): 319–325.

Grün, O. (2004b). *Taming Giant Projects: Management of Multi-Organization Enterprises*. Berlin: Springer.

Gulick, L. (1937). Notes on the theory of organization. In: *Papers on the Science of Administration* (eds. L. Gulick and L. Urwick), 1–46. New York: Harcourt.

Hales, C. (1986). What do managers do? A critical review of the evidence. *Journal of Management Studies* 23 (1): 88–115.

Hall, G. and Howell, S. (1985). The experience curve from the economist's perspective. *Strategic Management Journal* 6: 197–212.

Hamel, G. (2002). *Leading the Revolution*. New York: Plume.

Harris, F. and McCaffer, R. (2006). *Modern Construction Management*. Oxford: Blackwell.

Harty, C. (2005). Innovation in construction: a sociology of technology approach. *Building Research and Information* 33 (6): 512–522.

Harty, C. (2008). Implementing innovation in construction: contexts, relative boundedness and actor-network theory. *Construction Management and Economics* 26 (11): 1029–1041.

Hassan, S., McCaffer, R., and Thorpe, T. (1999). Emerging Clients' Needs for Large Scale Engineering Projects. *Engineering, Construction and Architectural Management* 6 (1): 21–29.

Henderson, R. and Clark, K. (1990). Architectural innovation: the reconfiguration of existing product technologies and the failure of established firms. *Administrative Science Quarterly* 35 (1): 9–30.

Hirsch, W. (1952). Manufacturing progress functions. *Review of Economics and Statistics* 34: 143–155.

Hofstede, G. (1984). *Culture's Consequences: International Differences in Work-Related Values*. Newbury Park: Sage.

Hofstede, G. and Hofstede, G. (2005). *Cultures and Organizations: Software of the Mind*. New York: McGraw-Hill.

Hollander, S. (1965). *The Sources of Increased Efficiency: A Study of DuPont Rayon Manufacturing Plants*. Cambridge: MIT-Press.

House, R., Hanges, P., Javidan, M. et al. (2004). *Culture, Leadership, and Organizations: The GLOBE Study of 62 Societies*. Thousand Oaks: Sage.

Janis, I. (1972). *Victims of Groupthink: A Psychological Study of Foreign Policy Decisions and Fiascos*. Boston: Houghton and Mifflin.

Jensen, M. and Meckling, W. (1976). Theory of the firm: managerial behavior, agency costs and ownership structure. *Journal of Financial Economics* 3 (4): 305–360.

Johnson, J. and Tatum, C. (1993). Technology in marine construction firms. *ASCE Journal of Construction Engineering and Management* 119 (1): 148–162.

Kahneman, D. (2011). *Thinking, Fast and Slow*. London: Penguin Books.

Kieser, A. and Nicolai, A. (2005). Success factor research – overcoming the trade-off between rigor and relevance. *Journal of Management Inquiry* 14 (3): 275–279.

Koontz, H. and O'Donnell, C. (1955). *Principles of Management: An Analysis of Management Functions*. New York: McGraw-Hill.

Koskela, L. and Vrijhoef, R. (2001). Is the current theory of construction a hindrance to innovation? *Building Research and Information* 29 (3): 197–207.

Krill, A. and Eibl, S. (1999). Tragverhalten der geneigten Elastomerlager des Bang-Na Expressway. *Bauingenieur* 74: 183–187.

Kyriazis, E. (2007). Frequency of Communication Within NPD Projects: Implications for Key Measures of success. In: *Proceedings of the Australian and New Zealand Marketing Academy Conference* (eds. M. Thyne, K. Deans and J. Gnoth), 872–879. Dunedin: Australian and New Zealand Marketing Academy.

Laborde, M. and Sanvido, V. (1994). Introducing new process technologies into construction companies. *Journal of Construction Engineering and Management* 120 (3): 488–508.

Lakatos, I. (1970). The methodology of scientific research programmes. In: *Criticism and the Growth of Knowledge* (eds. I. Lakatos and A. Musgrave), 91–196. Cambridge: Cambridge University Press.

Langford, D. and Male, S. (2001). *Strategic Management in Construction*. Osney Mead: Blackwell Science.

Laurent, A. (1983). The cultural diversity of Western conceptions of management. *International Studies of Management and Organization* 13 (1): 75–96.

Lewin, K. (1946). Action research and minority problems. *Journal of Social Issues* 2 (4): 34–46.

Li, H. and Love, P. (1998). Developing a theory of construction problem solving. *Construction Management and Economics* 16 (6): 721–727.

Lieberman, M. (1984). The learning curve and pricing in the chemical processing industries. *RAND Journal of Economics* 15: 213–228.

Lieberman, M. (1987). The learning curve, diffusion, and competitive strategy. *Strategic Management Journal* 8: 441–452.

Loosemore, M. (2014). *Innovation Strategy and Risk in Construction: Turning Serendipity into Capability*. Abingdon: Routledge.

Lu, S. and Sexton, M. (2006). Innovation in small construction knowledge-intensive professional service firms: a case study of an architectural practice. *Construction Management and Economics* 24 (12): 1269–1282.

Luhmann, N. (1971). Sinn als Grundbegriff der Soziologie. In: *Theorie der Gesellschaft oder Sozialtechnologie* (eds. J. Habermas and N. Luhmann), 25–100. Frankfurt a.M.: Suhrkamp.

Luhmann, N. (2013). *Introduction to Systems Theory*. Cambridge: Polity.

Luhmann, N. (2017). *Trust and Power*. Cambridge: Polity.

Manley, K. (2006). The innovation competence of repeat public sector clients in the Australian construction industry. *Construction Management and Economics* 24 (12): 1295–1304.

March, J. (1988). *Decisions and Organizations*. Cambridge: Basil Blackwell.

Masterman, J. (2002). *Introduction to Building Procurement Systems*. London: Routledge.

McKelvey, B. (1975). Guidelines for the empirical classification of organizations. *Administrative Science Quarterly* 20 (4): 509–521.

McSweeney, B. (2002). Hofstede's model of National Cultural Differences and Their Consequences: a triumph of faith – a failure of analysis. *Human Relations* 55 (1): 89–118.

Milgrom, P. (1989). Auctions and bidding: a primer. *Journal of Economic Perspectives* 3 (3): 3–22.

Miles, M. and Huberman, A. (1994). *Qualitative Data Analysis*. Thousand Oaks: Sage.

Miller, R. and Lessard, D. (eds.) (2000). *Strategic Management of Large Scale Engineering Projects: Shaping, Institutions, Risks, and Governance*. Harvard: MIT Press.

Mintzberg, H. (1973). *The Nature of Managerial Work*. New York: Harper and Row.

Mintzberg, H. (1980). Structures in 5's: a synthesis of the research on organization design. *Management Science* 26 (3): 322–341.

Mintzberg, H. (1994). *The Rise and Fall of Strategic Planning*. New York: The Free Press.

Mintzberg, H. (2009). *Management*. San Francisco: Berrett-Koehler.

Mintzberg, H. and Waters, J. (1985). Of Strategies, deliberate and emergent. *Strategic Management Journal* 6 (3): 257–272.

Mubarak, S. (2010). *Construction Project Scheduling and Control*. Hoboken: Wiley.

Nam, C. and Tatum, C. (1997). Leaders and champions for construction innovation. *Construction Management and Economics* 15 (3): 259–270.

Nemet, G. (2006). Beyond the learning curve: factors influencing cost reductions in photovoltaics. *Energy Policies* 34: 3218–3232.

Nonaka, I. and Takeuchi, H. (1995). *The Knowledge Creating Company: How Japanese Companies Create the Dynamics of Innovation*. New York: Oxford University Press.

North, D. (1990). *Institutions, Institutional Change and Economic Performance*. Cambridge: Cambridge University Press.

Nussbaum, M. (2018). *The Monarchy of Fear*. New York: Simon and Schuster.

Organization for Economic Co-operation and Development (OECD) (2005). *Oslo Manual: Guidelines for Collecting and Interpreting Innovation Data*, 3e. Paris: OECD.

Ozorhon, B. (2013). Analysis of construction innovation process at project level. *Journal of Management in Engineering* 29 (4): 455–463.

Parsons, T. (1991). *The Social System*. London: Routledge.

Pennanen, A. and Koskela, L. (2005): Necessary and Unnecessary Complexity in Construction, Proceedings of the First International Conference on Complexity, Science and the Built Environment, University of Liverpool, UK.

Perrow, C. (1967). A framework for the comparative analysis of organizations. *American Sociological Review* 32 (2): 194–208.

Perrow, C. (1986). *Complex Organizations: A Critical Essay*. New York: McGraw-Hill.

Picot, A., Dietl, H., Franck, E. et al. (2015). *Organisation: Theorie und Praxis aus ökonomischer Sicht*. Stuttgart: Schäffer-Poeschel.

Podolny, W. and Muller, J. (1982). *Construction and Design of Prestressed Concrete Segmental Bridges*. New York: Wiley.

Polanyi, M. (1958). *Personal Knowledge. Towards a Post-Critical Philosophy*. Chicago: University of Chicago Press.

Popper, K. (1945). *The Open Society and its Enemies - Volume Two: Hegel and Marx*. London: Routledge, 2003.

Popper, K. (1999). *All Life Is Problem Solving*. Oxford: Psychology Press.

Popper, K. (2002). *Logic of Scientific Discovery*. London: Routledge.

Popper, K. and Eccles, J. (1977). *The Self and its Brain*. Milton Park: Routledge.

Prade, W. and Surbeck, H. (1998). Segment-Verlegegeräte für die Hochstraßen in Bangkok. *Bauingenieur* 73 (9): 366–380.

Project Management Institute (ed.) (2017). *A Guide to the Project Management Body of Knowledge, PMBOK Guide*. London: The Stationery Office.

Puddicombe, M. (2011). Novelty and technical complexity: critical constructs in capital projects. *Journal of Construction Engineering and Management* 138 (5): 613–620.

Reichstein, T., Salter, A., and Gann, D. (2005). Last among equals: a comparison of innovation in construction, services and manufacturing in the UK. *Construction Management and Economics* 23 (6): 631–644.

Rosenberg, N. (1990). Why do private firms do basic research (with their own money)? *Research Policy* 19 (2): 165–174.

Rosenthal, J. (2005). *Struck by Lightning - the Curious World of Probabilities*. Toronto: HarperCollins.

Sargut, G. and McGrath, R. (2011). Learning to live with complexity. *Harvard Business Review* 89 (9): 68–76.

Schotter, A. (1981). *The Economic Theory of Social Institutions*. Cambridge: Cambridge University Press.

Schulz von Thun, F. (1981). *Miteinander Reden 1: Störungen und Klärungen*. Reinbek: rororo.

Scott, R., Levitt, R., and Orr, R. (eds.) (2011). *Global Projects – Institutional and Political Challenges*. Cambridge: Cambridge University Press.

Sexton, M. and Barrett, P. (2003). A literature synthesis of innovation in small construction firms: insights, ambiguities and questions. *Construction Management and Economics* 21 (6): 613–622.

Shannon, C. and Weaver, W. (1949). *The Mathematical Theory of Communication*. Urbana Champaign: University of Illinois Press.

Simon, H. (1955). A behavioral model of rational choice. *Quarterly Journal of Economics* 69 (1) S. : 99–118.

Slaughter, S. (1993). Builders as sources of construction innovation. *Journal of Construction Engineering and Management* 119 (3): 532–549.

Slaughter, S. (1998). Models of construction innovation. *Journal of Construction Engineering and Management* 124 (3): 226–231.

Slaughter, S. and Shimizu, H. (2000). 'Clusters' of innovations in recent long span and multi-segmental bridges. *Construction Management and Economics* 18 (3): 269–280.

Slevin, D. and Pinto, J. (2004). An overview of behavioral issues in project management. In: *The Wiley Guide to Managing Projects* (eds. P. Morris and J. Pinto), 67–85. Hoboken: Wiley.

Smircich, L. and Morgan, G. (1982). Leadership: the management of meaning. *Journal of Applied Behavior Science* 18 (3): 257–273.

Smith, A. (1776). *An Inquiry into the Nature and Causes of the Wealth of Nations*. Oxford: Oxford University Press (2008).

Snow, D. and Benford, R. (1988). Ideology, Frame Resonance, and Participant Mobilization. *International Social Movement Research* 1 (1): 197–217.

Spradley, J. (1979). *The Ethnographic Interview.* Belmont: Wadsworth.

Stobaugh, R. and Townsend, P. (1975). Price forecasting and strategic planning: the case of petrochemicals. *Journal of Marketing Research* 12: 19–29.

Strauss, A. and Corbin, J. (1998). *Basics of Qualitative Research: Techniques and Procedures Developing Grounded Theory.* Thousand Oaks: Sage.

Stuart, T. (2002). Interorganizational technology. In: *Companion to Organizations* (ed. J. Baum), 621–641. Oxford: Blackwell.

Sullivan, G., Barthorpe, S., and Robbins, S. (2010). *Managing Construction Logistics.* Chichester: Wiley-Blackwell.

Tatum, C. (1987). Process of innovation in construction firm. *Journal of Construction Engineering and Management* 113 (4): 648–663.

Tatum, C. (1988). Classification system for construction technology. *Journal of Construction Engineering and Management* 114 (3): 344–363.

Taylor, J. (1995). *Economics.* Boston: Houghton Mifflin.

Taylor, J., Dossick, C., and Garvin, M. (2011). Meeting the burden of proof with case study research. *Journal of Construction Engineering and Management* 137 (4): 303–311.

Thaler, R. (2015). *Misbehaving - the Making of Behavioral Economics.* New York: Allan Lane.

Tijhuis, W. and Fellows, R. (2012). *Culture in International Construction.* Abingdon: Spon Press.

Tirole, J. (1999). Incomplete contracts: where do we stand? *Econometrica* 67 (4): 741–781.

Tolman, E. (1948). Cognitive maps in rats and men. *Psychological Review* 55 (4): 189–208.

Tuckman, B. (1965). Developmental sequence in small groups. *Psychological Bulletin* 65 (6): 384–399.

Tulacz, G. and Reina, P. (2018): ENR's 2018 Top 250 International Contractors: Markets Begin to Turn Around. Engineering News Record. https://www.enr.com/articles/45048-enrs-2018-top-250-international-contractors-markets-begin-to-turn-around, accessed April 2019.

Tushman, M. (1979). Work characteristics and subunit communication structure: a contingency analysis. *Administrative Science Quarterly* 24 (1): 82–98.

Tushman, M. and Nadler, D. (1978). Information processing as an integrated concept in organizational design. *Academy of Management Review* 3 (3): 613–624.

Tushman, M. and Nelson, R. (1990). Introduction: technology, organizations and innovation. *Administrative Science Quarterly* 35 (1): 1–8.

Uher, T. and Zantis, A. (2011). *Programming and Scheduling Techniques.* London: Spon Press.

United Nations (ed.) (2014). *World Urbanization Prospects, 2014 Revision, Highlights.* New York: United Nations.

Vainio, T. (2011): Building Renovation – a New Industry?" Proceedings Conference on Management and Innovation for a Sustainable Built Environment, Amsterdam.

Van de Ven, A. (1986). Central problems in the management of innovation. *Management Science* 32 (5): 570–607.

Van De Ven, A. and Delbecq, A. (1974). A task-contingent model of work-unit structure. *Administrative Science Quarterly* 19 (2): 183–195.

Van de Ven, A., Polley, D., and Garud, R. (2008). *The Innovation Journey.* Oxford: Oxford University Press.

Van den Bos, G. (ed.) (2007). *APA Dictionary of Psychology*. Washington: American Psychological Association.

Walker, D., Hampson, K., and Ashton, S. (2003). Developing an innovative culture through relationship-based procurement systems. In: *Procurement Strategies* (eds. D. Walker and K. Hampson), 236–257. Oxford: Blackwell.

Wallis, J. and North, D. (1986). Measuring the transaction sector in the American economy, 1870–1970. In: *Long-Term Factors in American Economic Growth* (eds. S. Engerman and R. Gallman). Chicago: University of Chicago Press.

Watzlawick, P., Beavin, H., and Jackson, D. (1967). *Pragmatics of Human Communication – A Study of Interactional Patterns, Pathologies, and Paradoxes*. New York: W. W. Norton and Company.

Weick, K. (1995). *Sensemaking in Organizations*. Thousand Oaks: Sage.

Whetten, D. (1989). What constitutes a theoretical contribution? *Academy of Management Review* 14 (4): 490–495.

Wilke, H. (2000). *Systemtheorie I: Grundlagen*. Stuttgart: Lucius and Lucius.

Williamson, O. (1967). Hierarchical control and optimum firm size. *Journal of Political Economy* 75 (2): 123–138.

Williamson, O. (1985). *The Economic Institutions of Capitalism: Firms, Markets, Relational Contracting*. New York: Free Press.

Williamson, O. (1991). Comparative economic organization: the analysis of discrete structural alternatives. *Administrative Science Quarterly* 36 (2): 269–296.

Winch, G. (1998). Zephyrs of creative destruction: understanding the management of innovation in construction. *Building Research and Information* 26 (5): 268–279.

Winch, G. (2003). How innovative is construction? Comparing aggregated data on construction innovation and other sectors – a case of apples and pears. *Construction Management and Economics* 21 (6): 651–654.

Woodhuysen, J. and Abbey, I. (2004). *Why Is the Construction Industry So Backward?* Chichester: Wiley.

Woodward, J. (1965). *Industrial Organization: Theory and Practice*. London: Oxford University Press.

World Bank (2015): http://data.worldbank.org/indicator/FP.CPI.TOTL.ZG/countries/TH?display=default, accessed July 2015.

Wright, T. (1936). Factors affecting the costs of airplanes. *Journal of Aeronautical Sciences* 3: 122–128.

Yelle, L. (1979). The learning curve: historical review and comprehensive survey. *Decision Sciences* 10: 302–328.

Yin, R. (2013). *Case Study Research: Design and Methods*. Thousand Oaks: Sage.

Zaltman, G., Duncan, R., and Holbek, J. (1973). *Innovations and Organizations*. New York: Wiley.

Index

Advanced Construction Project Management: The Complexity of Megaprojects,
First Edition. Christian Brockmann.
© 2021 John Wiley & Sons Ltd. Published 2021 by John Wiley & Sons Ltd.